Optics

FOR

DUMMIES®

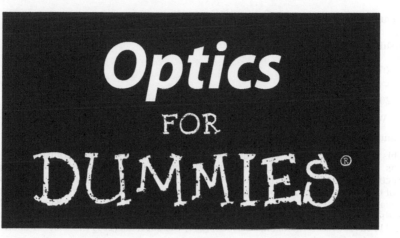

Optics

FOR

DUMMIES®

by Galen Duree, Jr., PhD

Wiley Publishing, Inc.

Optics For Dummies®

Published by
Wiley Publishing, Inc.
111 River St.
Hoboken, NJ 07030-5774
www.wiley.com

About the Author

Galen Duree, Jr., earned his PhD in physics, working on numerous laser systems and optical phenomena at the University of Arkansas–Fayetteville. He is presently a professor of physics and optical engineering at Rose-Hulman Institute of Technology in Terre Haute, Indiana. He is also the current director for the Center for Applied Optics Studies, which brings student teams together with companies looking for optical solutions. He runs the Ultrashort Pulse Laser Laboratory at Rose-Hulman, providing research opportunities to students of all disciplines at all levels (yes, even freshmen!) in many different areas.

Duree has worked with and for the Navy in the areas of high-energy laser systems, night vision, electro-optics, and ultrashort pulse laser applications. He has also worked with EG&G Technical Services in Crane, Indiana, in the areas of electro-optics and night vision systems. He currently consults with the Navy and SAIC, Inc., on high-power laser systems and ultrashort pulse laser applications.

When not working with students on research projects in optics, Duree loves spending time outside with his kids, working in the gardens, and working on his beloved 1979 Celica GT that still takes him to and from campus.

About the Contributors

Doug Davis is a licensed professional engineer working in the aerospace industry and as a consultant in his spare time.

Andrew Zimmerman Jones is the Physics Guide for About.com. He's studied and written about physics since 1993. Andrew holds a Physics degree from Wabash College with honors and awards, and he's the author of *String Theory For Dummies*.

Dedication

To my wonderful parents, Galen Sr. and Leslie Duree, for always supporting my inquisitive nature and helping me learn more about this wonderful world we live in.

This book is also dedicated to all those who have ever wondered if there is anything more to optics than eyeglasses and telescopes.

Author's Acknowledgments

This book is the result of many people's efforts, and I wish to thank them all. First, to all the students at Northwest Nazarene College and Rose-Hulman who have endured my lectures and asked a zillion questions over the years as I learned how to teach. I thank my acquisitions editor, Erin Calligan Mooney, and my project editor, Alissa Schwipps, for their patience and help in making this book possible. I also want to thank Andrew Zimmerman Jones and Douglas Davis for helping me with some elements as I started to run into a major time crunch. They have done a nice job, and I appreciate their input.

Last, but certainly not least, I want to thank my wonderful wife, Amber. By putting up with my discussions of picoseconds and explanations of things that probably only a physicist could appreciate over these many years, she helped me improve my understanding of things as well as refined my explanation of them. Without her initial and constant encouragement for writing this book and assistance with all the things going on at home and around home, this book would never have been written. I would like to thank my kids, Galen[3], Catherine, and Annalisa for their patience as they had to put up with the "Not now! Dad needs to finish his book!" line far too many times. I would also like to thank Annalisa for being my Little Person Editor, making sure that I used the right words and put the commas in the right places.

Publisher's Acknowledgments

We're proud of this book; please send us your comments at `http://dummies.custhelp.com`. For other comments, please contact our Customer Care Department within the U.S. at 877-762-2974, outside the U.S. at 317-572-3993, or fax 317-572-4002.

Some of the people who helped bring this book to market include the following:

Acquisitions, Editorial, and Media Development

Senior Project Editor: Alissa Schwipps

Acquisitions Editor: Erin Calligan Mooney

Senior Copy Editor: Danielle Voirol

Copy Editor: Megan Knoll

Assistant Editor: David Lutton

Technical Editors: Jason J.B. Harlow, PhD, Orven Swenson, PhD

Editorial Manager: Christine Meloy Beck

Editorial Assistants: Rachelle S. Amick, Alexa Koschier

Art Coordinator: Alicia B. South

Cover Photo: © iStockphoto.com / Henrik Jonsson

Cartoons: Rich Tennant (`www.the5thwave.com`)

Composition Services

Project Coordinator: Nikki Gee

Layout and Graphics: Carrie A. Cesavice, Nikki Gately, Joyce Haughey, Melissa K. Smith, Corrie Socolovitch, Christin Swinford

Proofreaders: Melissa D. Buddendeck, Melissa Cossell

Indexer: BIM Indexing & Proofreading Services

Publishing and Editorial for Consumer Dummies

Diane Graves Steele, Vice President and Publisher, Consumer Dummies

Kristin Ferguson-Wagstaffe, Product Development Director, Consumer Dummies

Ensley Eikenburg, Associate Publisher, Travel

Kelly Regan, Editorial Director, Travel

Publishing for Technology Dummies

Andy Cummings, Vice President and Publisher, Dummies Technology/General User

Composition Services

Debbie Stailey, Director of Composition Services

Contents at a Glance

Table of Contents

Introduction

*O*ptics is the study of light. This field of study includes finding out light's properties, investigating how light interacts with things (including itself), and figuring out how to make things that use light to send information or make measurements. As scientists and engineers learn more about light, they're developing new technologies that allow people to do new and exciting things.

Optics can be quite mysterious. Light is something that you probably take for granted. By it, you get much information about the world around you. It is always there, whether you pay attention to it or not.

But how do you get information from the light? Why don't you see images everywhere? How can light form images one instant and drill through a steel plate the next? *Optics For Dummies* takes the large body of optical phenomena around you and breaks it into small pieces to explain how and why optical events happen. With this knowledge, you can see how the properties of light make measurements and images and perform other tasks such as cutting or drilling.

As new optical technologies come to a store near you (whether it's in the form of 3-D television, a new data-storage device, or a new way to sense temperature in a room), this book gives you the basic understanding to help you figure out how these applications of light work.

Optics For Dummies is designed to help you understand the different optical phenomena, avoid common mistakes students make, and look at some of the basic design features involved in making practical devices. In addition to presenting typical information covered in optics classes, the material goes a little farther to provide hints of capabilities with light that may lead to your own significant understanding or invention involving light (and I hope it does).

About This Book

Optics For Dummies is written for you, dear optics student. It isn't an operator's manual or an optics textbook where you can get lost in fancy derivations or convoluted explanations. After many years of dealing with students' questions about optics, I've made note of the explanations that have been the most beneficial and worked hard to reduce the physics-y jargon and to concentrate on plain-English explanations.

If you aren't an optics student but simply have an interest in learning more about this field of physics, you can also benefit from this book. It's written to remove much of the mystery behind this incredibly useful science to help you appreciate what optics can do for you, too.

The great thing about this book is that *you* decide where to start and what to read. It's a reference you can jump into and out of at will. Just head to the table of contents or the index to find the information you want.

Conventions Used in This Book

I use the following conventions throughout the text for consistency and clarity:

- ✔ I format new terms in *italics* and follow them closely with an easy-to-understand definition.
- ✔ I also use italics to denote a variable (and its magnitude value) in text.
- ✔ **Bold** text highlights the action parts of numbered steps as well as the keywords in bulleted lists.

Also, as with many technical fields, many different conventions can be used to present equations and topics. I have adopted the notation that most textbooks and professional and scientific journals use when talking about optics. This way, when you come across a topic in your textbook or magazine article, this book can fit right in and provide you with a simple explanation about how to use an equation or describe a particular optical phenomenon. Here are some of the optics-specific conventions I use:

- ✔ **Metric units:** Because optics is really a subset of physics, I use metric units in the examples and equations. One notable exception: With lenses, I do some calculations with U.S. customary units.
- ✔ **Significant digits:** The examples presented in this book involve three significant digits. This convention allows me to just concentrate on working through an example instead of complicating matters by dealing with significant digits.
- ✔ **Definition of *light*:** In this book, I apply the term *light* to all electromagnetic radiation, not just to the electromagnetic waves your eyes can detect (as most people do). All electromagnetic radiation has the same properties, so what you observe about visible light happens with radio waves, x-rays, and any electromagnetic radiation between and beyond. The only thing that differs is the scale of the arrangements needed to see the different effects.

What You're Not to Read

Although I encourage you to read everything, I use a couple of indicators to flag material that you don't have to read. Technical Stuff icons indicate more-detailed or historical information related to a particular topic that isn't vital to understanding that topic. You can read these paragraphs if you want more information, but if you skip them, you won't miss anything that you need to understand the rest of the section.

Sidebars (text in gray boxes) look at particular applications or historical aspects of the topic at hand. You can read sidebars to find out about a specific situation where some optics phenomenon appears in the environment around you, discover a particularly useful application of a phenomenon, or find out more about a topic's evolution. Again, you can skip the sidebars without compromising your understanding of the rest of the text.

Foolish Assumptions

As I wrote this book, I made a few assumptions about you, the reader:

✔ You have a vague idea about light and its wave and particle properties and how to manipulate where light goes. If you need a briefing on this core info to refresh your memory, Part I summarizes the basic stuff.

✔ You're comfortable with algebra and a little matrix algebra for polarization. (Chapter 2 provides a review of all the math you need for this book.)

✔ You've had an introductory high-school or collegiate physics course and are familiar with solving problems and dealing with certain basic phenomena like mechanical waves.

If you have no physics background and picked up this book in an attempt to figure out what optics is, no worries! I present the material in a straightforward fashion, building on basic optics principles (covered in Part I) so that even without a physics background, you can still get the basic idea about a wide variety of optics phenomena and applications.

How This Book Is Organized

Optics is a very large field of study that touches many applications in many disciplines. To help you grasp the concepts and applications in manageable bites, the subject is split into several parts and smaller chapters.

Part I: Getting Up to Speed on Optics Fundamentals

If you don't know much about light, this part helps you quickly get a grasp of the main aspects of light that help you build things and, you know, see. If you are familiar with light, you can use this part to refresh your memory about things you may have forgotten.

Part II: Geometrical Optics: Working With More Than One Ray

This part looks at the particle property of light, which causes light to follow straight-line paths (called *rays*) between surfaces. This part is where you find out about making images and changing the properties of images using reflection or refraction of light.

Part III: Physical Optics: Using the Light Wave

This part deals with the wave properties of light. Here you explore optical polarization, the fundamental property used for optical data in fiber-optic networks and interference (which isn't a bad thing in optics). After you understand how interference is created, you're then ready to see how it can be used to measure optical properties as well as the dimensions of very small features on materials.

Part IV: Optical Instrumentation: Putting Light to Practical Use

This part deals with creating useful optical devices by manipulating the properties of light. You see how basic devices such as eyeglasses, microscopes, telescopes, and projectors work and check out some of the basic current and future light sources, including lasers.

A lot of modern technology relies on information carried by fiber-optic networks, so in this part, I also explain the basic elements of a fiber-optic link.

Part V: Hybrids: Exploring More-Complicated Optical Systems

This part deals with more-complicated optical systems that include two or more optical properties. I also cover simple optical devices, such as lasers, cameras, medical imaging equipment, night vision systems, thermal vision, speed guns, and telescopes, that address needs in particular applications.

Part VI: More Than Just Images: Getting Into Advanced Optics

Part VI looks at some rather interesting and complicated aspects of light. Things begin to work much differently when the amount of light sent into a material is very large or when you deal with a single photon or a couple of photons at a time. This part gives you a glimpse of some state-of-the-art research in optics — in particular, nonlinear optics and quantum optics — with some applications to show why you should care.

Part VII: The Part of Tens

Part VII provides ten simple experiments that you can do to see optics at work and to begin to give you some personal experience with light. It also covers ten historical experiments that helped improve the understanding of light and optical systems and introduces the people who made them possible.

Icons Used in This Book

Some information presented in this book is tagged with icons to help you identify important concepts and points to keep in mind when working on problems. Other icons indicate information that you don't need unless you want more details about the topic presented.

This icon indicates important points to keep in mind when dealing with concepts or equations to make sure that you get the correct answer.

A Tip icon highlights information that helps you deal with optical situations more quickly or easily.

When information goes into a little more detail than you may need about a concept or shows a higher-level application, I mark it with this icon.

The Warning icon flags information that highlights dangers to your solution technique or a common misstep that optics students may make.

Where to Go from Here

This book is set up to allow you to start anywhere you want. You can start at the beginning to get an idea about light, or you can go straight to the information you want about optics, properties of light, or applications. For example, turn Chapter 19 to find out how to build a telescope, or flip to Chapter 15 to find out how a fiber-optic link works. Optics is a highly diversified field of study, with applications from medicine to particle physics, and you don't have to read this book from cover to cover to understand the topics presented here. If you just want to find out about a certain topic, you can look up the subject of interest to you in the index and get the answers you need.

Optics is an exciting and very broad area of study, so however you choose to partake of the material, I hope you find an appreciation for all the different ways that you use and rely on light. Enjoy!

Part I
Getting Up to Speed on Optics Fundamentals

The 5th Wave   By Rich Tennant

"Paul, turn off your flashlight. There's a real interesting star cluster I'm trying to get a picture of."

In this part . . .

Optics is the study of light, so Part I is designed to provide you with the basic properties of light and some of the mathematics you need in order to use the various equations in the rest of the book. I explain the basic wave properties and particle properties of light, the experiments that caused the change from one model to the other, and the remarkable discovery of photons. You also find out about the three ways to produce light and the three basic processes that you can use to make light go where you want it to go. All of optics is based on the properties and models presented in this part, so they form the basis for all the other phenomena and devices you discover in this book.

Chapter 1

Introducing Optics, the Science of Light

· ·

· ·

*L*ight is probably one of those things that you take for granted, kind of like gravity. You don't know what it is or where it comes from, but it's always there when you need it. Your sight depends on light, and the information you get about your environment comes from information carried by the light that enters your eye.

Humans have spent centuries studying light, yet it remains something of a mystery. We do know many properties of light and how to use them to our benefit, but we don't yet know everything. Therefore, *optics* is the continuing study of light, from how you make it to what it is and what you can do with it. In fact, optics consists of three fields: geometrical optics, physical optics, and quantum optics. As we learn more about light, we find new ways to use it to improve our lives. This chapter shines a little, well, light on light.

Illuminating the Properties of Light

Because of an accidental mathematical discovery, light is called an *electromagnetic wave,* a distinction indicating that light waves are made up of electric and magnetic fields. You're probably used to thinking of light as the stuff your eyes can detect. For many people who work with light on a regular basis, however, the term *light* applies to all electromagnetic radiation, anything from ultra-low frequencies to radio frequencies to gamma rays.

Light has both wave and particle properties (as I discuss in the chapters in Part I), but you can't see both at the same time. Regardless of the properties, light is produced by atoms and accelerating charges. You can choose from many different arrangements to produce light with the desired wavelength or frequency (basically, the color that you want). Optics covers every light source from light bulbs to radio transmissions.

You have three ways to manipulate where light goes (that is, to make light do what you want): reflection, refraction, and diffraction, which I introduce in the following sections. You can use some basic equations to calculate the result of light undergoing all these processes. Optics then goes farther to investigate ways to find practical uses of these phenomena, including forming an image and sending digital data down a fiber.

Creating images with the particle property of light

You most commonly see the particle property of light when you're working with *geometrical optics,* or making images (see Part II). In this theory, the particles of light follow straight-line paths from the source to the next surface. This idea leads to the simplest type of imaging: shadows. Shadows don't give you a lot of information, but you can still tell the shape of the object as well as where the light source is.

Two important concepts in geometrical optics are reflection and refraction. *Reflection* describes light bouncing off a surface. *Refraction* deals with the bending of the path of the light as the light goes from one material to another. You can use these processes to create and modify images, and knowledge of these effects can also help you deal with factors called *aberrations,* which cause an image to be blurry. You can also use the lenses and mirrors that work with refraction and reflection to eliminate the washed-out effect you sometimes get when creating an image; if you have too much light, all the images created wash each other out, so all you see is light.

Harnessing interference and diffraction with the wave property of light

Physical optics, which I cover in Part III, looks at the wave properties of light. *Interference* (where two or more waves interact with each other) and *diffraction* (the unusual behavior of waves to bend around an obstacle to fill the space behind it) are unique to waves.

To explain *optical interference* (interference between light waves), you need to know about optical polarization. *Optical polarization* describes the orientation of the plane that the light wave's electric field oscillates in. In optics, only the electric field matters in almost all interactions with matter, because the electric field can do work on charged particles and the magnetic field can't. Several devices can change the polarization state so that light can be used for many different applications, including lasers and optical encoding.

The wave property allows you to use interference to help measure many things, such as the index of refraction and surface feature height or irregularities. Specifically, several optical setups called *interferometers* use interference for measurement.

Diffraction, the other unique wave phenomenon, determines resolution, which is how close two objects can be while still being distinguishable. Arrangements with many slits placed very close together create a *diffraction grating,* which you can use to help identify materials by separating the different colors of light the materials emit.

Using Optics to Your Advantage: Basic Applications

Understanding the basic properties of light is one thing, but being able to do something practical with them is another. (Head to the earlier section "Illuminating the Properties of Light" for more on these basic characteristics.) Putting the fundamental knowledge to good use means developing optical instruments for a wide variety of uses, as I discuss in Part IV. Here's just a taste of some of the practical applications of optical devices:

- **Manipulating images:** As I note earlier in the chapter, knowing how images are made and changed with different types of lenses or mirrors allows you to design simple optical devices to change what the images look like. Eyeglasses are designed and built to correct nearsightedness or farsightedness, and a simple magnifying glass creates an enlarged image of rather small objects.

  Physical characteristics limit how large an image a simple magnifier can make, so you can build a simple microscope with two lenses placed in the right positions to provide greater enlargement of even smaller objects. To see things far away, you can build a telescope and to project an image onto a large screen, you can build a projector.

- ✔ **Developing lighting:** You can use also use optics principles to design lighting sources for particular applications, such as specific task lighting, general area lighting, and decorative lighting. The development of incandescent light bulbs, compact fluorescent bulbs, and future devices such as light emitting diodes (LEDs) all start with knowledge of the optical properties of materials.

- ✔ **Seeing where the eye can't see:** Optics, and particularly fiber optics, can send light into areas that aren't directly in your line of sight, such as inside a collapsed building or a body. Fiber optics relies on knowledge of total internal reflection (see Chapter 4) to be able to trap light inside a small glass thread.

Expanding Your Understanding of Optics

The fundamental principles of optics can tell you what will happen with light in different situations, but making something useful with these principles isn't so easy. Applications of optics, including optical systems, combine two or more optical phenomena to create a desired output. Most applications of optics require knowledge about how optical principles work together in one system; making optical systems requires careful thought to make sure that the light behaves in the way you want it to when you look at the final result, whether that's light from a particular source (such as a light bulb or laser) or an image from a telescope or camera.

Why are such advanced applications important? Seeing how all the optics phenomena work together (often in subtle ways) is the point of optical engineering. Knowing how light interacts with different materials and being able to read this information has led to advances in important fields such as medical imaging and fiber-optic communication networks.

Considering complicated applications

Some optics applications, such as those in Part V, require combinations of many different optics principles to make useful devices. Cameras that record images require knowledge of image formation, focusing, and intensity control to make nice pictures. Holography and three-dimensional movies put depth perception and diffraction gratings to work. Many medical-imaging techniques exploit the effects of light and how light carries information.

Lasers are a special light source with many uses. Because lasers are light, you have to understand how light works so that you can use them effectively and safely. Lasers today are involved in medical applications, various fabrication tasks, numerous quality control arrangements, optical storage discs such as CDs and DVDs, and a variety of military and law enforcement applications (but no laser guns yet).

Complex imaging devices can also allow you to see in low- or no-light situations. Thermal cameras create images based on temperature differences rather than the amount of reflected light. The age-old arrangement of looking at the heavens requires modifications of the simple telescope to overcome some of the limitations of using refracting optics.

Adding advanced optics

Advanced optics (see Part VI) covers phenomena that aren't simply based on simple refraction. When the index of refraction — normally independent of the intensity of the light — changes with the intensity, weird things can happen, such as frequency conversion in crystals. The area of advanced optics that studies these effects is called *nonlinear optics,* and it has provided numerous new diagnostic capabilities and laser wavelengths.

Another area of advanced optics is single photon applications. Single photon applications show some rather bizarre behavior associated with the fact that light is in an indeterminate state unless you make a measurement. This subject (also presented in Part VI) is the basis for new applications in secure communications and super-fast computing.

Paving the Way: Contributions to Optics

The field of optics is full of contributions from students challenging the establishment and the established way of thinking. Part VII includes some experiments you can try to experience some of the optics principles presented in this book; building some simple optical devices lets you begin to discover the challenges of building optical systems. After all, experiments are the root of discovery, so Part VII also looks at some important optics breakthroughs and the people who performed them.

All this information allows you to see how knowledge of optics advanced with contributions from newcomers to the field as well as established optical scientists. Using the basic principles outlined in this book, you'll have enough knowledge to be able to delve deeper into any optics subject you encounter in school, work, or just your curiosity. As optics technology progresses, you'll have the basic background knowledge to tackle any of the technology paths that develop. After all, the field of optics benefited from contributions from many different levels; if that doesn't motivate you to make the next significant contribution to the science of light, I don't know what will.

Chapter 2

Brushing Up on Optics-Related Math and Physics

In This Chapter

▶ Cementing important algebra and trig concepts

▶ Catching some important wave physics topics

*H*opefully, you find this chapter refreshing, literally. I cover some math fundamentals, including basic algebra, trigonometry and inverse functions, and vectors and matrices, as they relate to optics. I provide a quick overview of the physics of waves (mechanical waves, wavefronts, and the wave function).

Working with Physical Measurements

Most of the notation and computation in optics are very similar to what you cover in an introductory physics course. However, a couple of units used in some optics calculations bear special mention.

Optics calculations that involve lenses sometimes use a unit of measure called a *diopter* (D) to rate the light-bending power of the lens. Diopters are typically used on eyeglass or contact lens prescriptions because they're a convenient choice: One diopter is equal to one inverse meter, and because most image location calculations depend on the inverse of a lens's focal length, using diopters means avoiding the use of inverse fractions.

Another unit worth noting is the decibel (dB). This scale has the following standard form:

$$\beta = 10 \log_{10} \frac{I}{I_0}$$

In this equation,

- β is the decibel value.
- I is the amount of light measured at a particular point.
- I_0 is the reference value to compare I to.

The decibel scale is used to give a relative value. In optics, it provides a measure of the amount of light detected at the output of a system relative to the amount of light put into the system.

Refreshing Your Mathematics Memory

Certain optics principles need more-basic math to help explain them properly. And because you've probably slept since you took your basic algebra and trigonometry courses (at least, I certainly hope you have), this section helps bring some of the pertinent math stuff back to life.

Juggling variables with algebra

Algebra is basically writing equations with variables and then manipulating those equations to find the values you need. Algebra follows a few basic rules, including the following:

- Dividing by zero is illegal, and the result is undefined (not infinity).
- You have to perform any action on both sides of the equal sign.
- Following the order of operations, especially with parentheses, is vital.
- You need to watch your signs really, really carefully.

And here are some algebra laws and definitions:

- **Commutative Law:** $a + b = b + a$ and $ab = ba$
- **Associative Law:** $a + (b + c) = (a + b) + c$ and $a(bc) = (ab)c$
- **Distributive Law:** $a(b + c) = ab + ac$
- **Additive Identity:** $0 + a = a$
- **Multiplicative Identity:** $1a = a$
- **Additive Inverse:** $a - a = a + (-a) = 0$
- **Multiplicative Inverse:** $\frac{a}{a} = a\left(\frac{1}{a}\right) = 1$

But of course, optics requires a bit more algebra than just knowing basic rules and laws. You actually have to be able to move variables around in some cases to find the missing value.

Look at Snell's law, which shows what happens when a ray of light is incident on a surface, as shown in Figure 2-1. A ray of light P travels through the first medium (air) and hits the surface of the second medium (glass) at point O, with an incident angle θ_1 measured relative to the surface normal — a line perpendicular to the surface. Snell's law says that the refracted ray Q will leave the surface with a refraction angle θ_2 measured relative to the normal. The relationship between the refraction angle, the two indexes of refraction, and the incident angle is described by the following equation:

$$(\sin \theta_1)n_1 = (\sin \theta_2)n_2$$

In this equation, θ_1 is the incident angle, θ_2 is the refracted angle, n_1 is the index of refraction for the first medium, and n_2 is the index of refraction for the second medium.

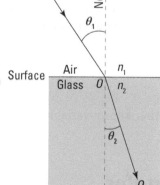

Figure 2-1: Refraction of light as it travels from one medium to another.

Using a little algebra, you can rewrite the equation to solve for some of these variables. Say you know everything about this example but the value of n_2, the second index of refraction. You can solve for that value by dividing both sides of the equation by the sin θ_2; with a little reorganization of the equation, you end up with

$$n_2 = n_1\left(\frac{\sin(\theta_1)}{\sin(\theta_2)}\right)$$

Now you can plug in the known quantities and get your answer.

Another example of useful algebra in optics has to do with the two-slit diffraction pattern. In high school physics class, you may have seen a demonstration where your teacher projected light through a very small slit; the light then passed through two more narrow slits parallel and equidistant from the first and then fell onto a screen some distance away, creating a pattern of light and dark vertical lines on the screen. What you were seeing on the screen is called an *interference pattern*, which I discuss later in the chapter and in Chapter 11.

Relative to that experiment, the equation for determining the phase difference between the two waves reaching a point on the screen is

$$x = \frac{2\pi a \sin(y)}{\lambda}$$

In this equation,

- ✔ a is the distance between the two slits.
- ✔ y is the angle between the waves and a plane through the slit perpendicular to the screen.
- ✔ λ is the wavelength of the light source.

Suppose you have the distance and ray angle and need to know the wavelength that creates a desired phase angle. You would manipulate the equation to solve for λ:

$$\lambda = \frac{2\pi a \sin(y)}{x}$$

Again, all you need to do is plug in the other known variable values and you have your answer.

Finding lengths and angles with trigonometry

Trigonometry is a portion of mathematics based on the special properties of the right triangle. You can use trigonometric functions to describe certain types of motion as well as to resolve unknown quantities from other known quantities.

The relationship between the sides of a triangle, ($h^2 = x^2 + y^2$), is known as the *Pythagorean theorem*. You can derive a number of the most common trigonometric functions from this theorem. I don't actually derive them in this section, but I do introduce some common functions and show you ways to use them.

Trigonometry functions use angles, which can have units of either degrees or radians. Both degrees and radians are dimensionless (unitless) numbers; they're both ratios of arc lengths, but they're measured differently. A full circle is divided into 360 degrees or 2π radians. Two full circles have 720 degrees or 4π radians.

You need to know which mode (degrees or radians) your calculator is operating in. Almost any program or calculator capable of performing trigonometry functions accepts the angle in either radians or degrees, and the wrong setting can have a significant impact on your calculations. For instance, the sine of 30 degrees equals 0.5, but the sine of 30 radians equals –0.988. Most calculators have a "rad" or "deg" setting you can change depending on what you want to work in.

Finding sides with trig functions

The six basic trigonometry functions are all a function of the right triangle. Consider the triangle in Figure 2-2.

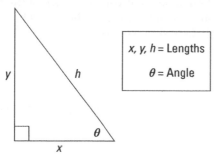

Figure 2-2: Example of a right triangle with features labeled.

x, y, h = Lengths
θ = Angle

The following list shows you the trigonometry functions commonly used and their relation to the triangle in Figure 2-2.

- ✔ **Sine:** $\sin \theta = \dfrac{y}{h} = \dfrac{\text{opposite}}{\text{hypotenuse}}$
- ✔ **Cosine:** $\cos \theta = \dfrac{x}{h} = \dfrac{\text{adjacent}}{\text{hypotenuse}}$
- ✔ **Tangent:** $\tan \theta = \dfrac{y}{x} = \dfrac{\text{opposite}}{\text{adjacent}}$
- ✔ **Cotangent:** $\text{cotangent } \theta = \dfrac{x}{y} = \dfrac{\text{adjacent}}{\text{opposite}}$
- ✔ **Secant:** $\text{secant } \theta = \dfrac{h}{x} = \dfrac{\text{hypotenuse}}{\text{adjacent}}$
- ✔ **Cosecant:** $\text{cosecant } \theta = \dfrac{h}{y} = \dfrac{\text{hypotenuse}}{\text{opposite}}$

The following other relationships and properties are also useful in using trig functions:

$$\tan \theta = \frac{\sin \theta}{\cos \theta}$$

$$\sin(-\theta) = -\sin(\theta)$$

$$\cos(-\theta) = \cos(\theta)$$

$$\tan(-\theta) = -\tan(\theta)$$

Finding angles with inverse trig functions

Another powerful function is the *inverse function.* You can indicate inverse functions in a couple of ways. One is to use the prefix *arc* in front of the trig function to denote "inverse." Another, perhaps more common indication (as shown on your scientific calculator), is with a superscript negative one, like this: $\sin^{-1}$ or $\cos^{-1}$. The inverse function of the same function leaves you with the original function variable: $\arcsin(\sin(\theta)) = \theta$. However, you can't mix function types (the $\arctan(\sin(x))$ doesn't equal x). In terms of a right triangle, the inverse functions allow you to find an angle if you know two of the triangle's sides. For instance, to find the angle θ in Figure 2-2, where you know the lengths of the legs of the triangle (x and y), you use

$$\theta = \tan^{-1}\left(\frac{y}{x}\right)$$

If you knew the length of the hypotenuse, h, and the leg, x, you'd use

$$\theta = \cos^{-1}\left(\frac{x}{h}\right)$$

Keep in mind that these functions, like most trigonometry functions, are only valid over certain ranges of values; you can't take the arcsine or arccosine of a number greater than one.

Consider Snell's law, which I describe in the earlier section "Juggling variables with algebra." You know the initial angle of incidence and the two indexes of refraction, but you need to find out what angle the ray will take after it's inside the second medium. To do so, you perform a few algebraic operations on the equation to get the following:

$$\sin(\theta_2) = \frac{\left(\sin(\theta_1)\right)n_1}{n_2}$$

And you think, "Wow, that's nice." But wait! You want to know what the angle is, not the sine of the angle. And you can determine just that by taking the inverse sine (arcsine or $\sin^{-1}$) of the sine function. By definition, the $\arcsin(\sin(\theta)) = \theta$, but you must be sure to use the inverse of the original function; as I note earlier, you can't mix function types. You apply the arcsin function appropriately to get

$$\arcsin\left(\sin\left(\theta_2\right)\right) = \arcsin\left(\frac{\left(\sin\left(\theta_1\right)\right)n_1}{n_2}\right)$$

Because the arcsine of a sine is the angle, the following equation gives you the answer you're looking for.

$$\theta_2 = \arcsin\left(\frac{\left(\sin\left(\theta_1\right)\right)n_1}{n_2}\right)$$

These inverse functions (arcsine, $\cos^{-1}$, arccot, and so on) are usually on your calculator somewhere; you may just have to do a little looking for them.

Exploring the unknown with basic matrix algebra

If you're rusty on linear algebra, including matrix algebra, getting reacquainted is worthwhile. Although the whole subject is much larger than I can present here, the following sections take a look at the vector and matrix and review some of the basics.

One of the most useful aspects of matrix algebra is how you can utilize it to help solve systems of simultaneous linear equations. Many problems, such as the ones in the preceding section, have more than one variable. If a problem has the same number of equations describing the system as it has variables, the problem usually has a unique solution. After you have written these equations in the appropriate form, you can enter them into matrix form and solve them by using matrix algebra.

Looking at matrix setup

A *matrix* is a set of numbers or values laid out in rows and columns. It's two dimensional and can have any number of *elements* (entries) within certain limits. (Matrix size eventually is limited by the ability to solve or process the elements.) The following matrix is a 4 x 4 matrix consisting of four rows and four columns.

$$\mathbf{Z} = \begin{bmatrix} a_{11} & a_{12} & a_{13} & a_{14} \\ a_{21} & a_{22} & a_{23} & a_{24} \\ a_{31} & a_{32} & a_{33} & a_{34} \\ a_{41} & a_{42} & a_{43} & a_{44} \end{bmatrix}$$

Each position in the matrix can be identified by a_{mn}, where m identifies the row and n identifies the column (a_{32} refers to the element in row 3, column 2). The preceding matrix is a *square matrix,* meaning it has the same number of rows as columns, but matrices don't have to be square. Notice also that the bold uppercase letter $\mathbf{Z}$ represents the matrix, and the lowercase letter a represents the entries.

A *vector* is a one-dimensional matrix and may look like this:

$$\mathbf{h} = \begin{bmatrix} a_1 & a_2 & a_3 & a_4 \end{bmatrix}$$

A vector (sometimes called an *array*) can also be referred to as a *row* or *column matrix.* The vector is denoted by a lowercase bold letter. A *scalar* is a matrix of only one row and one column (that is, one entry) and is usually denoted by italicized lowercase letters.

As you may have figured out, matrix dimensions are in rows by columns. The following $\mathbf{Z}$ matrix is a 3 x 2 matrix, and the $\mathbf{K}$ matrix is a 2 x 3 matrix.

$$\mathbf{Z} = \begin{bmatrix} a_{11} & a_{12} \\ a_{21} & a_{22} \\ a_{31} & a_{32} \end{bmatrix}$$

$$\mathbf{K} = \begin{bmatrix} a_{11} & a_{12} & a_{13} \\ a_{21} & a_{22} & a_{23} \end{bmatrix}$$

The letters used to name matrices above are arbitrary. Using letters that mean something to you may make sense, but the choice is yours. You may find that the identity matrix is commonly labeled $\mathbf{I}$, but again, that's not a hard-and-fast rule. (I discuss this special matrix in the later section "Solving systems of equations with matrices.")

Multiplying matrices

Optics uses some pretty complex equations consisting of several variables, and you frequently have to solve for more than one unknown variable. To help solve these complex equations for the variables of interest, you put them into matrix form and then use matrix math to solve them. One of the most important matrix math functions is matrix multiplication, but it's a little more complicated than it first looks.

Start with the most basic multiplication of a matrix. Consider the following matrix **P**:

$$\mathbf{P} = \begin{bmatrix} 4 & 2 & 0 \\ 5 & 3 & 1 \end{bmatrix}$$

If you multiply a scalar times **P**, you get

$$\mathbf{P}' = 2 \times \mathbf{P} = \begin{bmatrix} 2 \times 4 & 2 \times 2 & 2 \times 0 \\ 2 \times 5 & 2 \times 3 & 2 \times 1 \end{bmatrix}$$

The results are

$$\mathbf{P}' = \begin{bmatrix} 8 & 4 & 0 \\ 10 & 6 & 2 \end{bmatrix}$$

Multiplying one matrix by another matrix is a little more complicated but not difficult. For you to be able to multiply two matrices together, they must follow certain rules.

The first matrix in a multiplication must have the same number of columns as the second matrix has rows. The resulting matrix has as many rows as the first matrix and as many columns as the second matrix.

You can't multiply the following vectors because the row vector has fewer columns than the column vector has rows.

$$\begin{bmatrix} 1 & 7 & 5 \end{bmatrix} \begin{bmatrix} 2 \\ 4 \\ 1 \\ 6 \end{bmatrix}$$

The product of a row vector multiplied by a column vector is a scalar. For example, multiplying the following vectors gives you $2(8) + 6(1) + 3(4) = 16 + 6 + 12 = 34$, a scalar.

$$\begin{bmatrix} 2 & 6 & 3 \end{bmatrix} \times \begin{bmatrix} 8 \\ 1 \\ 4 \end{bmatrix}$$

The exact same process applies to bigger matrices. Matrix **K** has three columns, and matrix **Z** has three rows, so they meet the requirement. ***Tip:*** Remember to order your matrices correctly; **ZK** doesn't equal **KZ**.

$$\mathbf{K} = \begin{bmatrix} 2 & 8 & -1 \\ 3 & 6 & 4 \end{bmatrix} \quad \mathbf{Z} = \begin{bmatrix} 1 & 7 \\ 9 & -2 \\ 6 & 3 \end{bmatrix}$$

For bigger matrix multiplication, I recommend treating the calculations for the new matrix on an individual element basis. The resulting matrix will be a 2 x 2 matrix called **R**. If you write out the multiplication process, it looks like this:

$$r_{11} = 2 \times 1 + 8 \times 9 + (-1)6 = 68$$
$$r_{12} = 2 \times 7 + 8(-2) + (-1)3 = -5$$
$$r_{21} = 3 \times 1 + 6 \times 9 + 4 \times 6 = 81$$
$$r_{22} = 3 \times 7 + 6(-2) + 4 \times 3 = 21$$

The answer to $\mathbf{K} \times \mathbf{Z} = \mathbf{R}$ looks like this:

$$\begin{bmatrix} 2 & 8 & -1 \\ 3 & 6 & 4 \end{bmatrix} \times \begin{bmatrix} 1 & 7 \\ 9 & -2 \\ 6 & 3 \end{bmatrix} = \begin{bmatrix} 68 & -5 \\ 81 & 21 \end{bmatrix}$$

Solving systems of equations with matrices

As I note earlier in the chapter, you can use matrix math to help solve systems of simultaneous linear equations. But to fully utilize matrix math, you need to be familiar with a couple of special matrices: the identity matrix and the inverse matrix. An *identity matrix* is a square matrix of any size that has zeros for all elements except for the diagonal from the upper left to the lower right:

$$\mathbf{I} = \begin{bmatrix} 1 & 0 & 0 & 0 \\ 0 & 1 & 0 & 0 \\ 0 & 0 & 1 & 0 \\ 0 & 0 & 0 & 1 \end{bmatrix}$$

The *inverse matrix* multiplied by the original matrix creates the identity matrix. For a given matrix $\mathbf{A}$ and its inverse $\mathbf{A}^{-1}$, $\mathbf{A}^{-1}\mathbf{A}$ equals $\mathbf{I}$, where $\mathbf{I}$ is the identity matrix.

You can solve for the inverse matrix $\mathbf{A}^{-1}$ in any number of ways. However, the process for doing so is long for anything but a 2 x 2 matrix; I don't go into detail on that topic here. (You can find more info on matrix math in *Algebra II For Dummies,* by Mary Jane Sterling [Wiley].)

When showing your work isn't important, you can use your calculator to give you the inverse matrix. Enter the matrix on your graphing calculator and hit the inverse button and voilà: the inverse matrix!

Imagine you have the following three equations representing some system of interest, where x_1, x_2, and x_3 are unknowns:

$$x_1 + 2x_2 + 2x_3 = 1$$
$$2x_1 + 2x_2 + 2x_3 = 2$$
$$2x_1 + 2x_2 + x_3 = 3$$

The first step is to put these equations into a matrix form as $\mathbf{Ax} = \mathbf{y}$. First, make sure similar terms fall in the same place in each equation; rearrange the terms if necessary. Then write your matrices. The numbers in the first matrix are the coefficients on each term, the second matrix includes the variables, and the third matrix gives the answers.

$$\begin{bmatrix} 1 & 2 & 2 \\ 2 & 2 & 2 \\ 2 & 2 & 1 \end{bmatrix} \times \begin{bmatrix} x_1 \\ x_2 \\ x_3 \end{bmatrix} = \begin{bmatrix} 1 \\ 2 \\ 3 \end{bmatrix}$$

The next step is to multiply both sides of the equation by the inverse of **A** (that is, A^{-1}) because by definition, that operation equals one. Your equation now looks like this: $A^{-1}Ax = A^{-1}y$. Because $A^{-1}A = 1$, you're left with $x = A^{-1}y$.

You can use the earlier calculator shortcut to help you find the answer matrix. Find the inverse matrix and multiply that by the coefficient matrix, and your calculator gives you the answer matrix.

Having applied the inverse matrix, you're left with

$$
\begin{bmatrix} x_1 \\ x_2 \\ x_3 \end{bmatrix} = \begin{bmatrix} -1 & 1 & 0 \\ 1 & -1.5 & 1 \\ 0 & 1 & -1 \end{bmatrix} \times \begin{bmatrix} 1 \\ 2 \\ 3 \end{bmatrix} = \begin{bmatrix} 1 \\ 1 \\ -1 \end{bmatrix}
$$

And now you have your answer: $x_1 = 1$, $x_2 = 1$ and $x_3 = -1$.

Reviewing Wave Physics

In this section, I review the concept of a simple traveling wave function. I also discuss some basic properties and variables of the wave.

The wave function: Understanding its features and variables

To mathematically describe the motion of a wave, you use the wave function. The simplest wave form is a *harmonic* or *sinusoidal wave,* and its function has the basic form of $y(x, t) = A \sin(kx + \omega t + z)$.

The wave equation describes the displacement of particles (or molecules) caused by the wave around the equilibrium point. From Figure 2-3, you can see how the solid curved line (the sine function) moves above and below the axis or the point of equilibrium.

The wave function has a few properties you should be familiar with:

 ✔ **Displacement:** The wave function $y(x, t)$ is the description of all the particles (or molecules) in a medium that experience the wave. For the earlier equation, y is the displacement away from equilibrium. The letter x represents the position (space) coordinate, and t is the time coordinate.

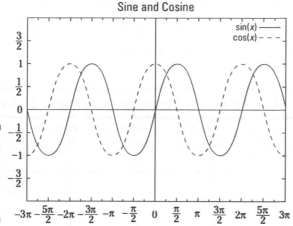

Figure 2-3:
A graph
of the sine
and cosine
functions.

✔ **Amplitude:** The letter *A* in the equation describes the *amplitude* of the wave. This value is the maximum magnitude (or height) the wave has as measured from the "at rest" position. When a body of water has no waves and isn't in motion, it's at rest. When something (such as a thrown rock) disturbs that water, the disturbance generates waves. As you look at the waves, you can't easily tell where the at-rest position would be because the clearest points of reference are the peak (crest) and the bottom (trough) of the wave. You may be tempted to measure the difference between these two features relative to the *y* direction and call that the amplitude, but that is actually 2*A*, or twice the amplitude.

✔ **Wave shape:** The next part of the equation, the sine function, is the mathematical description of the wave shape. Not all waves have the sine shape but can take on a number of different shapes that can be represented by other mathematical functions such as e^{ax}, cos, tan, and so on. Here's what the terms in parentheses (the sine function's *argument*) stand for:

  • **Number of waves:** The *k* stands for the *wave number* or *propagation number,* and it specifies the number of waves per unit distance. *kx* is 2π times the number of waves per unit distance (remember the argument of a sine function must be in radians or degrees) times a length.

  • **Number of oscillations:** ω is the angular frequency, and it specifies how many oscillations occur in a unit time interval, in radians per second. If you multiply that by time, *t,* you get a number in radians which varies with time. (***Note:*** To find the wave speed, *v,* you can divide the angular frequency by the propagation number: $v = \omega/k$.)

• **Phase:** z represents the initial phase. The initial phase of the wave is essentially an *offset variable,* which is the phase at time $t = 0$ and position $x = 0$. Table 2-1 presents values of the sine function at various positions of x and with different initial phases at $t = 0$.

Table 2-1	Values of a Sine Function with Phase Shifts *(z)*				
sin(x + z)	*Function at x = 0*	*Function at x = π/2*	*Function at x = π*	*Function at x = 3π/2*	*Function at x = 2π*
sin(x + 0)	0	1	0	−1	0
sin(x + π/2)	1	0	−1	0	1
sin(x + π)	0	−1	0	1	0
sin(x + 3π/2)	−1	0	1	0	−1
sin(x + 2π)	0	1	0	−1	0

Notice that the second equation at $x = 0$ has the same value as the first equation at $x = \pi/2$. The last equation at $x = 0$ has the same value as the first equation at $x = 2\pi$. This match is because the z component of the equation is "shifting" the function to the right, or leading the original base. If the z values were negative, the shift would be to the left, or trailing the base function.

By using z values equal to $\pi/2$, π, $3\pi/2$, and 2π, you get results that are integer values and allow you to see the repeating cycle of the function.

Medium matters: Working with mechanical waves

Mechanical waves are frequently described as disturbance in a medium (water or air) at rest. Examples of mechanical waves include waves on strings, water waves, and sound waves. You can further categorize mechanical waves based on how the wave moves particles as it travels through a medium:

✔ **Longitudinal waves:** Sound waves, for instance, are *longitudinal,* meaning the particles oscillate in the direction of the wave motion. Take a close look at a sound speaker. As the speaker cone is displaced, it pushes the air in the direction it's pointed.

✔ **Transverse waves:** Electromagnetic waves and string waves are considered *transverse waves.* The particles of the string wave oscillate in a

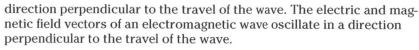

direction perpendicular to the travel of the wave. The electric and magnetic field vectors of an electromagnetic wave oscillate in a direction perpendicular to the travel of the wave.

Tie a jump rope to a door knob and snap the rope up and down to create a wave in the rope. The motion you use to create the wave (up and down) is perpendicular to the length of the rope. But the actual direction the wave travels is from your hand to the door knob.

In general, waves transport energy but not matter. Mechanical waves are a specific type of wave that isn't capable of transmitting energy through a vacuum. *Electromagnetic waves,* on the other hand, don't involve particles in a medium but rather deal with the fields associated with the electric and magnetic forces. Because electric and magnetic fields exist at any point in space, no medium is necessary to carry electromagnetic waves, and they can travel through a vacuum.

In each type of mechanical wave, the transmission of energy requires particles to be in contact with each other. In water or air, energy transmission requires that one particle push on the next, which pushes on the next. If a particle is missing, a void (or vacuum) ensues, and the one particle has nothing to push against. The particle on the other side of the void isn't pushed on, and the wave stops.

Using wavefronts in optics

Wavefronts and the wavefront diagram help you visually understand the motion and relationship of waves, whether the waves are incredibly small (measured in nanometers) like light waves or very large (tens of meters) like ocean waves. A *wavefront* is a line or surface that intersects all the waves in a collection of waves at the same value of the phase. A convenient phase value for mechanical waves is the crest or trough. So, a wavefront is the shape of the line you can draw that connects all the waves at their crests, for example. You can see the wave move by following the wavefront; all of the crests of the waves move together. You also can determine the wavelength of the waves by looking at the spacing between the wavefronts.

One of the simplest examples of a wavefront is visible in a clear container of water (the larger the better) at rest. Throw a rock or other small object into the water and watch the waves move out from the point where the object hit the water. Now imagine you can draw a line along the top of a single wave at the absolute highest point. Because the waves radiate out from the disturbance in a circle, you draw a circle. Now do the same for each subsequent wave you see. When you're done, you've drawn a number of concentric circles, each of which describes a wavefront. The line you draw along the highest point of the wave goes through all the particles that have the same phase, or same relative position on the wave. You can do the same thing at nearly any point of the wave (the trough, halfway down the face, and so on), and as long as you consistently choose the same points, you get the same effect.

You can use wavefronts to visually depict the motion of waves. From a wavefront diagram (see Figure 2-4), you can see a few important aspects, such as the following:

✔ The spacing of the waves

✔ How light is spreading as the size of the circles grows

✔ The energy or light density as the rays diverge (move away from each other)

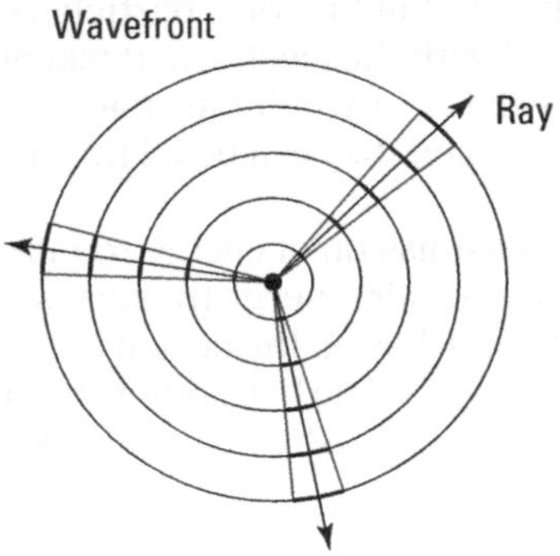

Figure 2-4:
Wavefront
diagram.

Chapter 3

A Little Light Study: Reviewing Light Basics

*L*ight is unlike anything else in your day-to-day experience. You can't touch it, walk on it, ride it to work, or, for the most part, feel it, but you rely on it when you check your reflection in the morning to see what your hair looks like (among myriad other tasks). Because light is so different than the other things in our world, isolating the properties of light to understand what light is has been difficult, even though people have studied light for centuries.

In this chapter, I review some of the major steps in the development of current understanding about light, which has allowed increasingly useful applications of light, such as CAT scans, laser surgery, and fiber-optic networks. I show some of the aspects of light necessary for the optics applications I present throughout this book. Finally, I list the three fields of study within optics that help narrow which properties of light are used in particular applications.

Developing Early Ideas about the Nature of Light

To understand how light works, you need to know what light is, and mankind has spent centuries trying to answer this question. I can't claim that I know exactly what light is, but I can tell you that you can look at light in two ways:

as a particle and as a wave. At first glance, these two theories for the nature of light seem to contradict each other. However, Albert Einstein was able to show that they aren't really all that different. (Flip to the later section "Einstein's Revolutionary Idea about Light: Quanta" for more on this connection.)

Pondering the particle theory of light

Although many earlier societies may have considered light, the ancient Greeks (about 500 BC) are typically credited with the earliest documented discussion about light. In their understanding, light was a particle of fire. They also thought sight worked like Superman's x-ray vision: Particles of light came out of the eye and interacted with the objects around them.

This sight theory is the origin of the idea that you can look into a person's eyes and tell something about his or her inner being, because the light particles came out of them (the inner fire). Expressions like "casting a glance" stem from this basic idea.

As time progressed, the idea of light as a particle remained, but where the light came from changed. Until the 1600s, light was considered a particle that objects emitted and the eye detected. The particles traveled in straight lines called *rays,* a concept that Isaac Newton quantified.

Newton was able to rigorously describe the phenomenon of *reflection* of light, where the light particles bounce off a surface, and the unique phenomenon of *refraction,* where the path of the light bends as it travels from one material to another. He was also able to show that white light contained many colors, and he developed the idea of *dispersion,* that the bending of the light path due to the refraction of light was dependent on the color of the light. At Newton's time, the particle theory of light could describe nearly all phenomena observed about light. But with all the successes of the particle theory of light, it wasn't able to explain everything people saw light do.

Walking through the wave theory of light

Because Newton was the man of the time, challenging the ideas he supported was very hard. Yet many young people (students, actually) looked at other ways to describe the properties of light. People such as Christiaan Huygens thought that light was actually a wave, like a wave on water or a string, rather than tiny particles, but these newcomers were ridiculed because this idea contradicted Newton. Huygens persisted, and soon the wave theory began to gain acceptance as scientists and engineers began finding new ways to look at what light could do.

Huygens used the idea of *wavefronts,* surfaces that connect waves at identical phase points (like a crest or a trough), to accurately model reflection and refraction of light. Unlike the particle theory, Huygens's wave idea was able to model *diffraction,* the ability of light to bend around an object into the space behind it. Thomas Young's famous two-slit experiment demonstrated diffraction and interference and was the first experimental determination of the wavelength of light.

Diffraction and interference are only possible with waves. Tiny particles, like very tiny spheres, can't diffract or interfere with each other. Over time, up through the early 1900s, light was considered a wave.

Taking a Closer Look at Light Waves

Particles are rather simple — hard little spheres that bounce around like balls. When it comes to describing a wave, the picture isn't so clear. When asked to describe a wave, you probably think of a wave on the surface of water. But water waves or waves on a stretched piece of string are *mechanical waves,* which require a medium — some kind of material — in order to exist. Mechanical waves can't exist in a vacuum, such as space, and light can. So what kind of wave is light? Read on!

If light is a wave, what's waving? Understanding electromagnetic radiation

Quite by accident, James Clerk Maxwell came up with an idea about what was waving in light waves: electric and magnetic fields. (See the sidebar "Waving hello to Maxwell's Wave equation" for more on this discovery.) This finding was quite unexpected, but useful; based on this equation, scientists could now test specific characteristics to see if these fields were indeed what light waves were. The other interesting breakthrough was that light waves didn't need a medium, an idea supported by a simple observation that if you stand outside in the sunlight, you can feel the warmth delivered to your body by light from the sun, which crosses 93 million miles in vacuum.

Based on Maxwell's serendipitous discovery, light is called an *electromagnetic wave.* Because what's waving in light waves is electric and magnetic fields, light is often referred to as *electromagnetic radiation.*

The electric field in light waves isn't like the electric field produced by separated, charged particles. Charged particles don't travel along with light waves. The electric field traveling in light is an *induced field,* like the electric field produced by a time varying magnetic flux. This is the principle by which credit card readers work.

According to Maxwell's wave equation, electromagnetic waves are *transverse waves,* meaning that the electric and magnetic fields oscillate perpendicular to the direction that the wave is traveling. The electric and magnetic fields behave the same, in that both can be described by sine functions and that the two fields are in phase. Translation: When the electric field reaches its maximum value, so does the magnetic field. You can write rather simple equations to describe the behavior of the fields:

$$E = E_{max} \sin\left(\frac{2\pi}{\lambda}z - 2\pi ft\right)$$

$$B = B_{max} \sin\left(\frac{2\pi}{\lambda}z - 2\pi ft\right)$$

In the equations,

- ✔ E represents the electric field in the wave.

- ✔ B represents the magnetic field in the wave.

- ✔ E_{max} represents the maximum value (amplitude) of the electric field.

- ✔ B_{max} represents the maximum value (amplitude) of the magnetic field.

- ✔ z represents the position along the path of the wave.

- ✔ t represents time.

- ✔ λ represents the wavelength of the wave.

- ✔ f represents the frequency of the wave.

Figure 3-1 illustrates a typical representation of an electromagnetic wave.

An interesting feature of this idea about light is that it's self-sustaining. Throughout the years, people have tried to come up with perpetual motion machines — devices that, once started in motion, continue forever. Because of thermodynamic principles, creating one of these devices really isn't possible. However, light appears to be the only phenomenon capable of perpetual motion. The two fields in electromagnetic radiation reinforce each other: The time varying electric field creates a time varying magnetic field, and the time varying magnetic field creates a time varying electric field. The fact that you can see the light that comes from stars that are many billions of light years away is evidence that light can travel for billions of years through space.

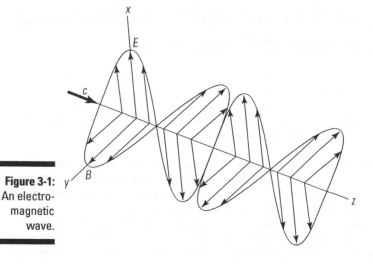

Figure 3-1:
An electro-
magnetic
wave.

Waving hello to Maxwell's wave equation

Scientists around the world performed experiments to find out what was waving in light waves. Because water waves require water and waves of strings require strings, scientists guessed that light waves required a special material. They called this material *ether* because it wasn't detectable by any means known. Albert Michelson and Edward Morely tried to detect the ether by using the Doppler effect. However, they didn't even detect a hint of this medium.

No one knew what was waving in light waves until a man named James Clerk Maxwell was looking at equations dealing with electricity and magnetism. He found that if he combined these equations in a particular fashion, he got a single mathematical equation that had the exact same form as the equation physicists and mathematicians used to model waves. One of the factors in this equation was the speed at which the wave travels, which came out to be very close to the measured speed at which light travels.

Remember: For light waves, the speed of light in vacuum is represented by c and is equal to 2.99792458×10^8 m/s. For most calculations you perform, 3.0×10^8 m/s works just fine, but you know how physicists like to be precise.

Dealing with wavelengths and frequency: The electromagnetic spectrum

The fields in electromagnetic radiation, in simplest terms, are described as a *sinusoidal wave*. Sinusoidal waves are *periodic,* which allows you to characterize the wave with standard wave parameters such as *wavelength* (the distance you move along the wave before you return to an identical value in the wave function) and *frequency* (the number of oscillations per second).

With mechanical waves, the speed, v, that the waves travel depends on the wavelength and the frequency, namely $v = f\lambda$. With electromagnetic waves, the speed in vacuum is equal to c, so $c = f\lambda$. This fact means that one term or the other completely specifies the wave; if you know the wavelength, you know the frequency, and vice versa. The range of observable values of wavelength and frequency is quite large, but visible light, the light that your eye can detect, occupies an extremely narrow portion of this range. Figure 3-2 shows the electromagnetic spectrum, with common names for the wavelengths or frequency.

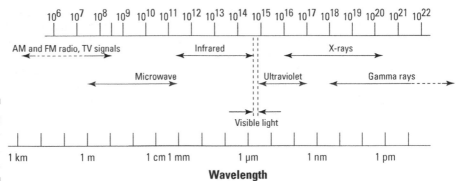

Figure 3-2: The electromagnetic spectrum.

Calculating the intensity and power of light

Waves, in general, propagate energy. If light is a wave, light must carry energy. A fundamental property of light's electric and magnetic fields is that they store energy (energy creates them). Using electromagnetic equations, you can find the energy stored in the electric and magnetic fields of electromagnetic radiation. Maxwell showed the following relationship between the maximum values of the field:

$$c = \frac{E_{max}}{B_{max}}$$

Because light waves travel in a particular direction, the *Poynting vector,* an equation based on electricity and magnetism theory, provides the rate at which light waves transport energy to a unit area of surface. The Poynting vector shows the direction that the wave travels in based on the orientation of the electric and magnetic fields. It also tells you the rate at which energy is delivered per unit of surface area. Because the fields follow a sinusoidal pattern, talking about the *intensity* or *irradiance,* the average power per area delivered by the wave is often useful, especially if the fields wiggle very quickly (like a million-billion times in a second). You calculate the intensity from the following equation:

$$I = \frac{1}{2} \frac{E_{max}^2}{\mu_0 c}$$

In this equation,

- ✔ I is the intensity of the electromagnetic wave. It has units of watts/square meter.

- ✔ E_{max} is the amplitude of the oscillating electric field.

- ✔ $\mu_0 c$ is the impedance of vacuum (a constant) and has a value of 377 ohms.

The intensity of the electromagnetic wave is a useful parameter for numerous applications, such as optical data transmission and laser machining, which I cover in many later chapters in this book.

Most introductory physics textbooks use the term *intensity,* but most optics texts call this parameter *irradiance.* Although I use *intensity* in this chapter to help you transition to the more formal term, later chapters use *irradiance.*

Einstein's Revolutionary Idea about Light: Quanta

For many decades, the wave picture of light seemed to be the proper model. All experimental observations, especially interference and diffraction, seemed pretty conclusive that light was a wave phenomenon. Although some still used the particle/ray picture in some circumstances, light was regarded as a wave.

As technology improved, scientists performed other experiments. By the beginning of the 20th century, scientists could determine the wavelength of light and then observed results that the wave theory couldn't explain. These results coincided with the beginning of the era of modern physics, which saw the beginning of quantum mechanics, and led to a new model for light. The following sections look at the evolution of modern light theory, including problems with the wave theory and Einstein's quanta breakthrough.

Uncovering the photoelectric effect and the problem with light waves

Scientists in the early 1900s began looking at a phenomenon called the *photoelectric effect*. Basically, they connected electrodes to a battery and placed them in an *evacuated* glass ball (meaning the air was removed from the ball). They shined light onto one of the electrodes, a large metal plate. Because light is a wave and so transports energy, the incident light gave energy to the electrons. When the electrons absorbed enough energy, they were ejected from the metal plate and traveled as part of the current in the electrical circuit connected to the electrodes. The results of this experiment showed four problems with the wave theory of light:

- ✔ **No electrons were emitted if the frequency of the light was below a certain value, no matter how bright the light was.** This frequency was called the *cutoff frequency,* and it shouldn't exist with light waves. According to the wave theory, electrons would just wait until they absorbed enough energy to leave the metal.

- ✔ **The maximum kinetic energy of the electrons was independent of the light intensity for incident light that was above the cutoff frequency.** Per the wave theory, greater intensity would mean more energy for the electrons, increasing their kinetic energy.

- ✔ **The maximum kinetic energy of the electrons increased with increasing frequency of the incident light.** Wave theory indicated that the maximum kinetic energy of the electrons should depend on the incident intensity, not the frequency of the light.

- ✔ **The electrons were ejected from the metal nearly instantaneously, even when the intensity was very low.** According to the wave theory, electrons should have to wait and absorb energy from the incident light waves.

This experiment was repeated many times, but the results never changed. The results of the photoelectric experiment had provided significant evidence that the wave model, although maybe not completely wrong, was, at least, incomplete.

Merging wave and particle properties: The photon

In 1905, Albert Einstein published a paper on electromagnetic radiation where he proposed a correction to the wave model of light. Ironically, his idea went back to the particle model for light (see "Pondering the particle theory of light" earlier in this chapter). Instead of making light a small, solid ball, Einstein reimagined the wave as a localized wave, which he called *quanta.* Later, the quantum of light were called *photons,* which is the name scientists use today.

I like to think of photons as wave packets. You can see a cartoon representation of a photon in Figure 3-3. In the figure, you can see the wiggly field inside, but it's confined to a small space. The wave theory of light, on the other hand, basically treated light waves as extending infinitely, which is why that theory failed to explain the photoelectric effect.

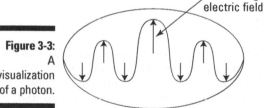

Oscillating electric field

Figure 3-3:
A
visualization
of a photon.

Einstein proposed another change to the wave theory: that the energy carried by each photon depends on the frequency of the light, calculated by the following:

$$E = hf$$

In this equation,

- ✔ *E* is the energy carried by the photon (*not* the electric field of the electromagnetic wave).
- ✔ *h* is Planck's constant and has a value of 6.626×10^{-34} J·s.
- ✔ *f* is the frequency of the light.

According to wave theory, waves continually transfer energy as long as they exist, meaning that the wave can transfer any amount of energy into, say, electrons in a metal. With Einstein's relatively simple frequency idea, changing the wave theory to the photon theory allowed the light to still have wave properties and particle properties and have a single picture for the nature of light that was consistent with experimental observations. Einstein's quanta idea allowed for a complete understanding of the observations made from many experimental situations, including those related to the photoelectric effect. The photon is the current model for light.

Let There Be Light: Understanding the Three Processes that Produce Light

Trying to understand the properties of light is important, but so is understanding the processes that produce light, especially when looking at how light and matter interact. You can attribute all sources of light to three basic processes: atomic transitions, accelerating charged particles, and matter-antimatter annihilation.

Atomic transitions

When you go into a dark room, you usually turn on a light to see what's in the room. The source of the light is the light bulb, but the light emitted by the bulb involves atomic transitions. Because an atom has basically two parts, electrons and the nucleus, atomic transitions cover processes in both parts.

Electronic transitions

According to the Bohr cartoon for the atom, electrons are bound to the positively charged nucleus. The paths of the electrons in the Bohr model form concentric circles around the nucleus. Instead of being traces of the actual paths of the electrons, they represent energy levels. Figure 3-4 shows this situation.

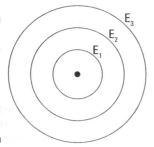

Figure 3-4:
The Bohr model of the atom with electron energy level transitions.

In order for an electron to move to a higher energy state, the electron must absorb energy equal to the difference in energy between the two energy states. The electron can gain this energy either by absorbing a photon of the right energy or via inelastic collisions that give the electron enough energy to change levels. The electron stays in a higher energy level, often referred to as an *excited state* because the electron has more energy for a certain amount of time; it then gives up its energy in the form of a photon, a process called *spontaneous emission,* so that the electron can move to a lower energy state. Although emitting of a photon isn't the only way an electron can give up energy, it's the one appropriate for this discussion.

Nuclear transitions

The nucleus has energy levels like the electrons bound to a nucleus do, and it can be put into an excited state. Once there, it wants (so to speak) to get to a lower energy state at some instant of time. An excited nucleus can get rid of its excess energy by emitting one of three things: an alpha particle, a beta particle, or a gamma ray (plus other fundamental particles such as neutrinos). Of these three, only the gamma ray is a form of light. Gamma rays have extremely short wavelengths and very large amounts of energy. Constructing situations where you can see the wave properties of gamma rays is difficult, so many people tend to think of them as particles, which is the only property left if you can't see the wave properties. When a nucleus goes to a lower energy state and emits a gamma ray, it's emitting light, so it's a another source of light, although not a source that you'd probably want to work with.

Accelerated charged particles

Accelerating charged particles emit light. A radio transmitter antenna uses this property to produce radio waves, which are a form of electromagnetic waves with a relatively long wavelength. The AC current produced by the electrical circuit (the transmitter, in this case) directly determines the frequency or wavelength of the radio waves. By varying the amplitude or frequency of radio waves, information is sent out over large areas so that you can hear what's going on through your favorite radio station.

Whenever charged particles are stopped quickly, like the electrons in a cathode ray tube, the particles emit electromagnetic radiation called *bremsstrahlung radiation.* When I was a child (way before plasma, LCD, or LED flat screens), my parents cautioned me not to stand right in front of the TV because I would hurt my eyes. I don't know whether they were aware of it, but because of the radiation emitted by the quickly decelerating electrons hitting the phosphor screen, I could've hurt more than just my eyes!

Charged particles that are moving in a circle with constant speed also emit electromagnetic radiation. You may recall from your physics class that any object that moves along a curved path must be accelerating, even if the speed is constant, because the direction of the velocity vector is changing. The radiation emitted by particles moving in a circle is called *synchrotron radiation*. Electrically neutral particles, like the neutron, don't give off light when they accelerate.

Matter-antimatter annihilation

Antimatter particles are rare, but when they appear, they look just like regular matter, except that their charge is different. A *positron* is an antiparticle that has the same mass as an electron but a different charge (a positive charge). As long as the positron doesn't encounter any electrons, it works just like any other free, positively charged particle. If it runs into an electron (which is a rare occurrence), the two annihilate each other and produce light in the form of a gamma ray.

Introducing the Three Fields of Study within Optics

Light is a sufficiently complicated entity that appears to have both wavelike and particlelike properties. The nice thing, though, is that any one experiment can observe only one property. This characteristic has led to the creation of three fields of study within optics: geometrical optics, physical optics, and quantum optics. Each field concentrates on one of the properties of light and allows you to design applications based on that one property.

Geometrical optics: Studying light as a collection of rays

Geometrical optics, the oldest field of optical study, concerns itself with the particle property of light. As I note in "Pondering the particle theory of light" earlier in the chapter, Newton proposed that light moves in rays until something that can change the direction (such as a surface) gets in the way. However, after the ray leaves the surface, it again continues in a straight line. By changing the surface or selecting materials with the right properties, you can use these surfaces to control where images appear. Geometrical optics primarily deals with the formation of images, which I talk about in Part II.

Physical optics: Exploring the wave property of light

Physical optics deals with the wave property of light, looking at the uniquely wavelike phenomena such as interference and diffraction and the concept of *polarization,* the orientation of the electric field in the light waves. Most data communication networks and, recently, some medical imaging techniques depend on research in this area of optics. I cover this aspect of light in Part III. You can put digital information on a light wave by working with the wave properties of light and the optical properties of a material. I discuss this idea in Chapter 21.

Quantum optics: Investigating small numbers of photons

Although geometrical and physical optics allow you to talk about the general behavior of light by considering the effects of many rays or waves, they aren't too helpful for taking on one ray (or photon) or wave by itself. *Quantum optics* looks at the fundamental nature of light: the nature of photons and the bizarre fact that photons have wavelike and particlelike properties at the same time. Studying small numbers of photons lets you see effects that are usually blurred out when you have large numbers of photons.

The results of quantum optics studies are leading to a new application in the area of secure communication. Head to Chapter 22 for more on the characteristics of single photons and possible applications.

Chapter 4

Understanding How to Direct Where Light Goes

In This Chapter

▶ Discovering the three basic processes that can effect where light travels

▶ Finding the models for each process

*I*n geometrical optics (see Chapter 3), light travels in straight-line paths called *rays*. If these straight paths were the only direction light traveled, you couldn't rely on sight to get around a room. Fortunately, three processes can change the direction that light travels: reflection, refraction, and diffraction. Not only can these processes change the direction that light travels, but they can also put other kinds of information, such as digital information that your computer makes, onto light. To be able to design systems that use these processes for specific applications, you need a way to predict how the system will change the direction of the light. Using the concept of rays provides a convenient way to gain insight into how these processes change where light goes.

In this chapter, I show you these basic processes and how they relate to the particle and wave properties of light that I discuss in Chapter 3. I show you the equations for reflection and refraction (which both particle and wave theories identically predict), explaining how they work with single rays. Diffraction is more complicated and can only happen with the wave property, so I explain what it is but leave the details for Chapter 13.

Reflection: Bouncing Light Off Surfaces

Reflection (light bouncing off a surface, much like a ball bouncing off the floor or a wall) is the simplest process for changing the direction that light travels in. Reflection is the primary way you get information about the world around you. Light travels from a source to an object, bounces off the object, and travels toward your eye, delivering the information about the object. To design an optical device that uses reflection requires that you be able to determine the orientation of the light and, with that information, predict where the light will go after bouncing off the surface. In this section, I show you the standard convention for determining the orientation of the incoming and outgoing (reflected) light when using the law of reflection.

Determining light's orientation

If you look at a single ray *incident* on a surface (traveling toward and hitting a surface), you need to be able to determine the ray's orientation relative to the surface so that you can calculate the new path that the light will take after bouncing off the surface. The standard convention used to describe the orientation of rays relative to a surface is to measure the angle that the ray makes relative to a line called the *surface normal* drawn perpendicular to the surface. Although using the surface rather than this constructed line may seem easier, the boundary conditions for electromagnetic radiation incident on a surface are much easier to work with when you use the surface normal.

With the convention for measuring the orientation of the light rays relative to a surface established, the law of reflection is

$$\theta_i = \theta_r$$

In this equation, θ_i is the *angle of incidence,* and it's the angle that the ray traveling toward the surface *(incident ray)* makes as measured from the surface normal. θ_r is the *angle of reflection,* or the angle that the ray is traveling away from the surface *(reflected ray)* after it has bounced off the surface. Figure 4-1 shows a general example of ray reflection and the measured angles.

Suppose you're looking at a single ray incident on the surface of a mirror. You measure the incident angle to be 48 degrees with respect to the surface normal. With this incident angle, you see from the law of reflection that the reflected ray leaves the mirror at an angle of 48 degrees in the opposite direction of the incident angle.

Figure 4-1:
The general geometry showing reflection of a single ray from a surface.

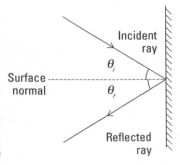

Surface normal

Incident ray

θ_i

θ_r

Reflected ray

Keep in mind that the reflection equation relates only to the angles' magnitudes (values) and not to their directions. The reflection angle's direction is technically the opposite of the incident angle's direction because light doesn't generally bounce back in the direction it arrived unless it's at normal incidence, which I cover later in the chapter. So the θ_r should have a minus sign to indicate the change in direction, but you can do the calculations without this sign in the equation as long as you remember to flip the angle of reflection in your diagram.

Understanding the role surface plays in specular and diffuse reflection

The law of reflection I cover in the preceding section works nicely for a single ray. However, most applications in optics require many more rays. A collection of rays exhibit two different types of reflection, specular and diffuse, depending on what the reflecting surface is like. Most optic elements (such as lenses or mirrors) in applications use a special set of rays called *parallel rays,* where many rays of light all travel in the same direction, to quantify what the elements will do with light. You can think of the light in this situation as several arrows traveling parallel to each other moving along a straight line. The way that the optic elements treat parallel rays is very important information to have if you want to design an optical device. Using this concept, the following sections delve farther into the two types of reflection.

Specular reflection

Specular reflection occurs whenever the surface is flat or smooth. You can consider a surface optically smooth if any surface feature height (sort of like mountains on a plain) is much smaller than the wavelength of the incident light. If parallel rays are incident on a smooth surface, the reflected rays remain parallel, as shown in Figure 4-2. An example of a surface that exhibits specular reflection is your bathroom mirror.

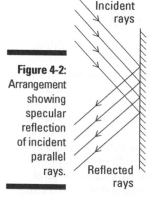

Figure 4-2:
Arrangement showing specular reflection of incident parallel rays.

Incident rays

Reflected rays

Diffuse reflection

Diffuse reflection occurs on rough surfaces — that is, if any surface feature height is on the order of the wavelength of the incident light or larger. If parallel rays are incident on a diffuse surface, the reflected rays bounce off that surface in random directions. The law of reflection applies to each ray separately, but because the surface is randomly oriented, the surface normal (determined at the point where the incident ray hits the surface) is randomly oriented for each ray. This situation causes the reflected rays to leave the surface in random directions, as shown in Figure 4-3.

Diffuse reflectors destroy the parallel orientation of the incident rays after reflection. An example of a surface that exhibits diffuse reflection is this page. You can see the words from any direction because the light that hits the page reflects in all directions. If the page were smooth, you'd only be able to see the image if you moved your eye into the path of the reflected rays.

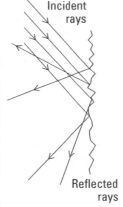

Figure 4-3:
Arrangement showing diffuse reflection of incident parallel rays.

Incident rays

Reflected rays

Appreciating the practical difference between reflection and scattering

In optics, you often hear two terms dealing with bouncing light off of things: reflection and scattering. Don't confuse them. *Reflection* occurs at a surface that is usually defined by something solid, such as glass or metal, but can also occur on the well-defined surface of liquids, such as a calm pond. Reflected light bounces off the surface and travels in the opposite general direction of the incident light. *Scattering* occurs as light travels through materials such as gasses or liquids with small particles (even atoms) that change their location constantly, but without a well-defined, static surface. Scattering sends light in all directions, much like diffuse reflection, except that reflection happens at a surface and can't send light in the forward direction (the same general direction that the incident light was traveling) like scattering does. Two kinds of scattering determine what colors you see the causes of the scattering as: Rayleigh and Mie. The difference between the two depends on the size of the particles causing the scattering relative to the wavelength of the light incident on the particles.

Rayleigh scattering

Rayleigh scattering occurs when the wavelength of the light is much, much larger than the diameter of the particles in the medium that the light is traveling through. Even though the wavelength is so large, the electrons in the particles can still respond to the electric field in the light. The effect is generally very weak, so you usually can't see Rayleigh scattering unless the light travels through a lot of material. The model for Rayleigh scattering is basically that the amount of incident light scattered by the particles is proportional to λ^{-4}. Therefore, light with a smaller wavelength scatters more than light with a longer wavelength.

You can see the effect of Rayleigh scattering every time you look at a blue sky. The sunlight is practically light with all visible wavelengths, from red to blue. When the sunlight enters the atmosphere, the oxygen and nitrogen molecules are on the order of 0.2 nanometers, which fulfills the requirement for Rayleigh scattering because 0.2 nanometers is much smaller than the 400-nanometer wavelength of blue light. Scattering is more likely to happen with shorter-wavelength light, so the blue light in the sunlight is scattered much more than the longer-wavelength red light. Because scattering occurs in all directions, the sky appears blue no matter what direction you look, as long as the sun isn't near the horizon (at sunset or sunrise).

Rayleigh scattering is weak, so not all blue light is scattered out of the sunlight; enough blue light is left to prevent objects from looking red in the daylight. When the sun is near the horizon, the light from the sun is more or less traveling directly toward you; the sunlight travels through a much larger thickness of air than it does when it's overhead, so more blue light is scattered out of the sunlight. Because the blue light is scattered out of this light as it goes through more of the atmosphere, the only light left is red (or orange). At sunrise and sunset, you see the leftover sunlight, after the blue has been taken out by scattering.

Mie scattering

Mie scattering involves particles that are about the size of the wavelength of the incident light or slightly larger. Unlike Rayleigh scattering, each particle in Mie scattering scatters all the incident light equally. Unlike reflection, which generally also works for all wavelengths, Mie scatters the light in all directions, even in the forward direction.

You can see the effects of Mie scattering when you look at fog, clouds, or a glass of milk. Milk appears white because it scatters all incident light in all directions equally. All wavelengths present means the stuff appears white. Fog appears gray simply because the density of water droplets is much less than the particles in milk. Clouds appear white in a blue sky because the ice crystals or water droplets scatter all incident light equally in all directions. The dark area in white clouds is just the parts of the cloud that are in the shadows, not a different scattering condition.

Refraction: Bending Light as It Goes Through a Surface

Refraction happens when the path that the light travels bends as it goes from one material into another. It can only occur in materials that are transparent to the incident light; a smooth sheet of metal reflects light because the light can't enter the metal, but a piece of glass refracts the light because some of the light goes through the surface.

Making light slow down: Determining the index of refraction

To calculate how much a light's path bends, you need to know something about how the light and material interact with each other. This interaction is called the *optical property* of the material. For refraction, the quantity you need is the *index of refraction,* which allows you to determine how the light interacts with the material. The basic definition of the index of refraction is

$$n = \frac{c}{v_{apparent}}$$

In this equation,

- ✔ n is the index of refraction of the material.
- ✔ c is the speed at which light travels in a vacuum: 3.0×10^8 m/s.
- ✔ $v_{apparent}$ is the apparent speed at which light travels through the material.

In a vacuum, light always travels at a speed of 3.0×10^8 m/s. A material, even a solid, is mostly nothing — that is, nothing makes up the space between the atoms. But in reality, light travels much more slowly through materials. If light travels 3.0×10^8 m/s through the nothingness between atoms, what causes the travel time to increase? Several theories for this discrepancy exist, but for the situations present in this book, the definition of the index of refraction works just fine, so I don't complicate the situation unnecessarily.

Calculating how much the refracted ray bends: Snell's law

If you know the index of refraction for the materials the light travels through (see the preceding section), you need one more piece of information to determine how refraction changes (bends) the path that the light follows in the new material. This element is the light ray's orientation relative to the surface the light is traveling toward. As I note earlier in the chapter, you measure this orientation relative to the surface normal, not the actual surface.

So with the index of refraction of the materials and the angle of the incident ray measured relative to the surface normal, you can determine the angle that the refracted (transmitted) ray is bent by using the law of refraction or Snell's law:

$$n_1 \sin(\theta_1) = n_2 \sin(\theta_2)$$

In this equation,

- ✔ n_1 is the index of refraction of the material that the light starts out in.
- ✔ n_2 is the index of refraction of the material that the light is traveling into.

  To help you keep the n factors straight, just remember that n_2 and "into" sound alike.

- ✔ θ_1 is the angle of incidence.
- ✔ θ_2 is the *angle of refraction,* the angle that the ray traveling away from the surface makes with the surface normal.

Using Snell's law, you can predict what will happen to the refracted ray in a qualitative fashion. You have only three possible situations:

- ✔ When light travels from a material with a relatively low index of refraction (compared to the index of refraction of the second material), the refracted ray bends toward the surface normal. This situation means that, if you do your calculations correctly, θ_1 will be larger than θ_2.
- ✔ When light travels from a material with a relatively high index of refraction (compared to the index of refraction of the second material), the refracted ray bends away from the surface normal. θ_2 is larger than θ_1.
- ✔ If the incident light is parallel to the surface normal ($\theta_1 = 0$ degrees, which is called *normal incidence*), the refracted ray doesn't bend. This fact means that for light at normal incidence, the path the light follows into the material doesn't change; the light continues along a straight line through the surface and into the second material.

Suppose you want to find the angle of refraction for a light ray that is incident from air on a glass surface with an angle of incidence of 35 degrees. Table 4-1 lists the indices of refraction for some common materials; the index of refraction for air is practically 1.00, and the one for a common glass is 1.50.

Table 4-1	Indices of Refraction for Common Materials
Material	*Index of Refraction (yellow light)*
Air	1.00
Water	1.33
Glass	1.50
Oil	1.20
Plastic	1.40

Using this information in Snell's law,

$$n_1 \sin (\theta_1) = n_2 \sin (\theta_2)$$
$$1.00 \sin (35°) = 1.50 \sin (\theta_2)$$

Solving for θ_2,

$$\theta_2 = \sin^{-1}\left(\frac{0.57358}{1.50}\right) = 22°$$

Because the air has a lower index of refraction than glass, notice that the refracted ray bends toward the surface normal; the angle θ_2 is smaller than the incident angle, θ_1.

Bouncing light back with refraction: Total internal reflection

A special case of refraction takes place when light travels from a high-index material to a lower-index material. If the angle of incidence is at the *critical angle*, the refraction angle is 90 degrees, meaning that the light doesn't enter the second material! If the light is incident at an angle greater than the critical angle, the light bounces back into the first material; the light behaves just as if reflected, but this bounce-back is a refraction phenomenon called *total internal reflection*. Unlike regular reflection, total internal reflection transmits no light into the second material.

The critical angle for total internal reflection depends on the index of refraction of the two materials and is given by

$$\theta_c = \sin^{-1}\left(\frac{n_2}{n_1}\right)$$

In this equation,

- θ_c is the critical angle for the two materials.
- n_1 is the index of refraction of the material the light starts out in.
- n_2 is the index of refraction of the material the light is traveling into.

Total internal reflection can only occur when the light starts out in a material with a higher index of refraction than the index of refraction of the material that the light is trying to get into.

Suppose you want to determine the critical angle for water and air so that you can tell whether you can see an object on the bottom of a swimming pool from where you're standing. From Table 4-1, the index of refraction for the water is 1.33, and the index of refraction for air is 1.00. You want to see whether light can escape from the water into the air so that you can see the object. Because total internal reflection can only occur when the light starts in the higher index material, n_1 is 1.33 and n_2 is 1.00. Using the total internal reflection expression, you get:

$$\theta_c = \sin^{-1}\left(\frac{1.00}{1.33}\right)$$

This math gives you a critical angle value of 48.8 degrees. For you to actually be able to see the object, you need to account for refraction with Snell's law (flip to the earlier section "Calculating how much the refracted ray bends: Snell's law") to determine where you need to put your eye. But if you're standing on the edge of a swimming pool and the object is at the bottom of the diving section, 12 feet below the surface and 20 feet away, you can't see the object because the light is incident on the surface at an angle of 59 degrees, which is greater than the critical angle; because of total internal reflection, the light never leaves the water.

One application of total internal reflection is fiber optics; check out Chapter 15 for more details.

Varying the refractive index with dispersion

Refraction is, in general, a complicated phenomenon. The index of refraction includes the effect of the atoms in the material, but because the atoms making up the materials are different, the index of refraction is different, as is the way the light interacts with the atoms. Placing atoms in large numbers makes this environment even more complicated because the many atoms treat light with different energies, or frequencies, differently.

Light is usually characterized by its wavelength or frequency (see Chapter 3) and not its color because the perception of color is too subjective. As the electromagnetic spectrum in Chapter 3 shows, light with larger energy has a higher frequency and smaller wavelength. In the electromagnetic spectrum, this combination is at the bluer or violet end of the visible spectrum. Appearing at the red end of the spectrum, light with lower energy has a lower frequency and longer wavelength.

As I note in "Making light slow down: Determining the index of refraction" earlier in the chapter, light takes longer to travel through a material than through a vacuum because the light interacts with the material's atoms such that it takes longer to get through the material. Light with less energy and lower frequency slows down less than light with more energy and higher frequency. This frequency-based difference in travel time through the material is called *dispersion*. The technical definition of dispersion is that the index of refraction depends on the wavelength (or frequency) of the light, but basically, dispersion is the fact that the index of refraction isn't the same for all colors of light.

In terms of the visible spectrum, red light (low energy) takes less time to travel through a material than blue light does. Of course, materials are more complicated than this picture may indicate, but you get the idea. In general, a material has a larger index of refraction for blue light ($v_{apparent}$ is smaller) than for red light. In terms of Snell's law, blue rays bend more than red rays in the same material. Blue and red light represent the extremes in the visible spectrum, but the index of refraction usually varies smoothly for the wavelength (colors) in between these two.

Dispersion occurs in all refractive (transparent) materials, and you must account for it when designing optical systems that use elements operating based on refraction. I briefly discuss a common situation involving lenses and dispersion in Chapter 7.

Birefringence: Working with two indices of refraction for the same wavelength

Aside from dispersion (see the preceding section), materials usually have the same index of refraction for all orientations of the electric field in the light. However, some materials, usually nearly pure crystals, have two indices of refraction for the same wavelength of light. This situation is called *birefringence* (bahy-ruh-*frin*-gince). The difference is due only to the orientation of the light's electric field. Therefore, light with its electric field oscillating in the vertical direction experiences a different index of refraction from the index of refraction for an electric field oscillating in the horizontal direction. This phenomenon has significant consequences when designing certain optical systems, as I discuss in Chapters 10 and 16. Birefringence is what allows you to put information on laser beams for fiber-optic communications, and I present the basic concept for this application in Chapter 21.

Diffraction: Bending Light around an Obstacle

Diffraction is a phenomenon that happens around you all the time; you're just not able to perceive it. *Diffraction* occurs when light encounters an obstacle, such as an edge of a piece of paper or a piece of metal with a hole in it, and expands to try to fill the area behind the obstacle. Figure 4-4 shows conceptually what happens when light encounters a piece of metal with a slit in it.

The light in Figure 4-4 initially is traveling straight down onto the slit in the metal. Think of the white areas as the crests of the light waves and the dark areas as the troughs of the light waves. After the light travels through the slit, it expands into the area behind the metal. Notice that the wave fronts are now curved and propagate along the directions indicated by the arrows at the bottom of the figure. Also notice that the brightness gradually changes from white to black as the light waves travel, indicating the formation of bright and dark areas after the slit. Refraction doesn't cause this bending or expansion; the light isn't moving from material to material, so refraction can't take place. This expansion can't be explained with the particle theory of light and is one reason why the wave theory of light had to be taken seriously. (Chapter 3 sheds more light on the development of light theories.)

Even though diffraction and refraction both deal with the bending of light (and they rhyme), they don't describe the same phenomenon. Refraction requires a change in index of refraction (see the earlier section "Refraction: Bending Light as it Goes Through a Surface"). Diffraction doesn't depend on the index of refraction and happens only when an opaque obstacle is present in the path of the light.

Diffraction occurs with any wave phenomena, such as water waves or sound waves. You can experience diffraction of sound if you and a friend stand on either side of a corner of a building with no signs, other buildings, and so on nearby to reflect the sound. If your friend talks in the direction parallel to the wall, you can still hear him even though you can't see him because the sound waves bend around the obstacle (the corner of the building) and get to your ear, as long as you aren't too close to the wall.

You can experience diffraction of sound fairly easily because its wavelength is rather long (on the order of centimeters). Visible light, on the other hand, has a wavelength that is about 100,000 times smaller than sound's, so you can't as easily construct situations where you can see the diffraction of light. For you to see the effects of diffraction with your eye, you need an obstacle that is small enough to make the diffraction dramatic enough. One way to create this effect is to make a small hole in a thick piece of paper. Shining light on the hole, you can see a diffraction pattern on a screen placed behind the paper. I present specific cases of diffraction in more detail in Chapter 13.

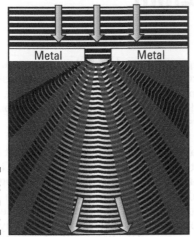

Figure 4-4: Diffraction in action.

Interference of light from the same source causes diffraction patterns; light from different sources causes interference patterns. You can tell which phenomenon is responsible for the light and dark pattern you see based on the spacing between the light and dark areas, or fringes. (*Fringes* result from the interference of waves; see Chapter 11.) Interference causes equally spaced fringes; diffraction causes fringes that aren't equally spaced, and I discuss this topic in more detail in Chapter 13.

Part II
Geometrical Optics: Working with More Than One Ray

The 5th Wave   By Rich Tennant

"Is that mirror working? I know I don't look like that."

Part II
Geometrical Optics:
Working with More
Than One Ray

In this part . . .

This part concentrates on the particle property of light, which is probably the oldest field of study in optics and the one you use on a day-to-day basis. Geometrical optics looks at the straight-line paths that light follows and is concerned, for the most part, with creating images. Using the particle property of light, I show you why you can see images in some situations and not in others. Through the concepts of reflection and refraction, I show you how you can create images and change their properties. The material presented in this part helps form the basis for the more complicated optical systems that you encounter later in this book.

Chapter 5

Forming Images with Multiple Rays of Light

The primary use of optics for the longest time has been to produce images. *Images* are representations of (or things that look like) objects that appear at locations other than where the actual objects are. Geometrical optics allows you to model image formation by using rays. You can alter the characteristics of the images by using the principles of refraction or reflection to make the image appear in a way that's most helpful to you. For instance, telescopes allow you to see distant objects clearly by enlarging the image of a distant planet; microscopes create enlarged images of tiny objects so that you can study microscopic organisms.

Reflection and refraction are essential, but they aren't the only properties necessary to create an image. Chapter 4 discusses the basic processes of reflection and refraction in terms of a single ray, but creating sufficiently detailed, recognizable images requires many more rays. Each ray that reflects off an object brings only a single piece of information about the object.

Think of an image as a jigsaw puzzle, where each piece is the small amount of information carried by a single ray. If you put one piece on the table, you probably can't tell what the picture is supposed to be (unless you cheat and look at the box). You have to have a large number of pieces put together in the correct way for the image to appear. The same principle applies to how rays of light produce an image. You can't see the information carried by one ray because it's too dim.

In this chapter, I show you the basic processes that create an image and which image characteristics you can affect by how you design the optical system. I present the primary tool used in designing optical imaging systems: the focal point and the focal length of lenses and mirrors.

The Simplest Method: Using Shadows to Create Images

The simplest image of an object is a *shadow,* made when an opaque object in the path of a light source prevents the light source from traveling through to the surface behind the object. Shadows are like many other images in some respects; to be recognizable, they require more than one ray. However, shadows aren't colorful (the shading that appears in a shadow has nothing to do with details in the middle of the object), and they don't show many details other than those features that are in the path of the light near the edge of the object.

The size of the shadow depends on the orientation of the object relative to the light. An object that's upright with a light source close to the ground creates a long shadow, whereas the same object creates a very small shadow if the light source is overhead. Geographers use this characteristic to map out stereographic maps of terrain with the sun as a light source.

Shadows are a little more complicated than this simple case because the characteristics of the shadow also depend to a large extent on the size of the light source or how close the object is to a small light source. Figure 5-1 illustrates the parts of a shadow, the umbra and penumbra.

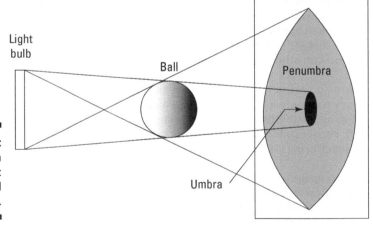

Light bulb

Ball

Penumbra

Figure 5-1:
Parts of a
shadow:
umbra and
penumbra.

Umbra

✔ **Umbra:** The darkest part of the shadow is the *umbra.* This region of the shadow is closest to the object. If you were to stand in the umbra, you would not be able to see the light source. This part of the shadow can take up the entire shadow when the light source is a *point source,* a light source that's very small and/or far away.

✔ **Penumbra:** The *penumbra* is the part of the shadow that isn't completely dark. This region of the shadow usually gets light from the parts of the light source that aren't blocked by the object, making the region appear gray and not as dark as the umbra. If the object is smaller than the light source and moved closer to the light source, light from other parts of the light source can cross over, hit the ground, and begin to brighten the shadow, making the shadow disappear. As the object moves farther away from the light source, the penumbra becomes darker until the whole shadow is taken up by the umbra. You can use these characteristics about shadows (umbra and penumbra) to tell how far an object is from a light source.

Shadows and eclipses

If you've paid attention to solar eclipses, you've probably noticed that only people in certain parts of the world can see the eclipses as total eclipses (where the sun is completely blocked out). This variation is due to the orientation of all the eclipse bodies relative to the light source. For a total solar eclipse, the sun completely disappears for only those places where the umbra of the moon falls on the earth. The other places see the penumbra of the moon's shadow fall on the earth, so the eclipse is partial, and part of the sun's disc is still visible.

Another special situation can happen with solar eclipses. Because of the varying distances of the sun and the moon from the earth, the moon may create an antumbra even when the centers of the sun and moon are directly aligned with the surface of the earth. (*Antumbra* is the astronomical term for a part of a shadow that exists when an object is viewed to be completely surrounded by the light source.) You can see the moon blocking the center part of the sun, but the moon isn't big enough (or close enough to the earth) to completely block out the sun. You see a ring, or *annulus,* around the edge of the moon. This situation is called an *annular solar eclipse.* You can see all these astronomical shadows in the following figure.

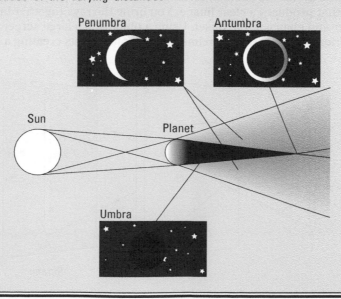

Forming Images Without a Lens: The Pinhole Camera Principle

Each ray of light that bounces off an object carries information about that object. If you can arrange all these rays properly, you can generate a clear image of the object. But when you look at the shadow of an object on a wall where a large collection of rays from that object hit, the image isn't clear. In fact, it's blurred out completely. If each ray carries a puzzle piece of information about the object, the image appears as though someone just dumped out the box of pieces; you see a random arrangement of the pieces, so you see nothing discernible. The trick is to get rid of the rays that don't help clarify the image.

If you trace the rays that bounce off an object and then hit a wall somewhere else, you can see that the rays hit random points on the wall; each image is slightly offset from all the other images deposited on the wall, such that the net effect is a washed-out image you can't see. However, if you restrict which rays hit the wall, you stop the appearance of numerous offset images and you improve your chances of seeing the image. You can imagine this situation with multiple projectors projecting the same image of an object. If all the projectors are pointed at the wall but slightly offset, you don't see a clear image. But if you turned off all but one projector, the image is clear.

The principle of a pinhole camera is to do just that: select rays that pass through a common point and block all of the other rays. This setup (shown in Figure 5-2) has the effect of turning off all the other projectors except for the one that produces a clear image. You bring the rays together in the proper order and eliminate the offset. Using a pinhole in a large screen prevents the addition of rays carrying an image that's offset, thus creating a clear image.

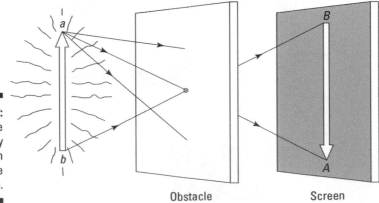

Figure 5-2:
The image created by a pinhole in an opaque screen.

Obstacle   Screen

You can try out the pinhole principle by poking a small hole in a large piece of cardboard. If you place the pinhole such that the screen (cardboard) is in a darkened room but looking out at a well-lit scene, you see the scene on the screen very clearly because the pinhole prevents all the light from the scene from hitting the screen and washing out the image.

Eyeing Basic Image Characteristics for Optical System Design

You can use many techniques to create an image. The previous sections introduce the basic idea about bringing rays together in the proper fashion so that you can make a clear image. But many applications require an image that has different properties than what you can make with a pinhole camera (see the preceding section). Making an image larger or smaller, for instance, requires manipulating the rays so that the image produced has the desired characteristics. In order to effectively create a useful image beyond using shadows or pinholes, however, you need to be aware of some important characteristics of images.

The type of image created: Real or virtual

Real images are ones you can see on a card or a screen. The light that makes the image actually passes through the plane where the image appears. The image produced by the pinhole camera is a real image. (See the earlier section "Forming Images Without a Lens: The Pinhole Camera Principle" for more on this image-forming technique.)

On the other hand, you can't see *virtual images* on a screen. The light that makes the image appears to originate from the image but doesn't actually pass through the plane where the image appears. These images typically require an additional method (such as the lens of your eye) to make them viewable.

The orientation of the image relative to the object

Orientation is usually based on the vertical direction, just because it's convenient and clear to draw. The orientation of an image is either erect or inverted, determined relative to the orientation of the actual object. If the image has the same vertical orientation as the object, the image is *erect*. If the image is upside-down relative to the object, the image is *inverted*.

The size of the image relative to the object

Many applications for optics require images that are different (especially in size) from how you observe the image of an object on the retina of your eye. Sometimes, you may want to reduce the object, such as a panoramic view of a mountain range, so that it fits on a reasonable piece of paper. Other times, you may want the image of an object to be enlarged, such as an amoeba under a microscope. To quantify and model the size change of an image, optical engineers work with numerous magnification factors. The most common magnification is *lateral magnification,* which is simply the measurement of a part of an object (its height, for instance) compared to the measurement of the same dimension in the image. The lateral magnificationis given by

$$M = \frac{h'}{h}$$

In this equation,

- ✔ M is the lateral magnification of the image.
- ✔ h' is the height of the image.
- ✔ h is the height of the object.

When you're designing an optical system, you often want to predict what the magnification of the image will be, but this prediction is hard to make if you don't have an image to work with. Geometrical optics to the rescue! In all imaging situations, geometry shows you a relationship between the magnification of the image and the ratio of the distances that both the object and the image appear relative to the optical element that's producing the image. The more useful form of the lateral magnification formula is

$$M = -\frac{i}{o}$$

In this equation,

- ✔ M is the lateral magnification of the image.
- ✔ i is the distance that the image appears from the optical element that's producing the image.
- ✔ o is the distance that the object is located from the optical element.

Using the proper sign convention for a particular optical element (lens or mirror), the magnification can also give you the orientation of the image relative to the object: If the magnification turns out to be a negative number, the image is inverted relative to the object. If the sign is positive, the image is erect.

Zeroing In on the Focal Point and Focal Length

In geometrical optics, you can use the process of refraction or reflection to produce images that are clear and bright enough to be seen. In order to design optical systems that are useful, engineers need to characterize the optics in some fashion so that the engineer can design the proper optics and determine the correct placement of the optics in the system.

For lenses and mirrors, the two basic optical elements in geometrical optics, optical engineers can perform a computer simulation called *ray tracing,* in which a computer program calculates the paths that many different rays travel through an optical element or system. However, an easier technique allows for a simple calculation to figure out where an image will appear and what its characteristics will be. All optical elements that form images use a special point, which I cover in the following sections, determined by what the element does with multiple rays of light.

Determining the focal point and length

Geometrical optics characterizes lenses and mirrors by their treatment of a special collection of rays called *parallel rays,* which, not surprisingly, are rays all parallel to each other. The lenses and mirrors that can change how the rays propagate need to have a curved surface. For simplicity, I talk about spherical element optics in this section. That is, the surfaces I discuss are curved like the surface of a ball but aren't round like a ball; they're like a section you'd cut off a ball.

To determine the effect of the curved surface on the collection of rays, you characterize the surfaces by the radius of the curved surface. For the optic calculations, the center of the sphere that the surface comes from is called the *center of curvature.* If you place a nail at the center of curvature and tie a string to the nail on one end and a pencil on the other, the pencil traces out the surface. The length of string, or the distance from the center of curvature to the surface, is called the *radius of curvature.*

Most curved mirrors or lenses either converge the parallel rays to a point or make them diverge in such a way that they seem to come from a point, called the *focal point.* (The radius of curvature determines how much the lens or mirror converges or diverges the parallel rays.) The distance that this focal point lies from the optical element (lens or mirror) is called the *focal length.*

In geometrical optics, the focal length of the optical element provides all the necessary information you need to figure out where the image will appear and what characteristics the image will have. See Chapters 6 and 7 for more on using the focal length for these elements.

Differentiating real and virtual focal points

Some optical elements cause incident parallel rays to converge to a point. Because the light actually passes through this point, the focal point is called a *real focal point.*

Other optic elements cause the incident parallel rays to diverge, but not in a random fashion. The rays diverge in such a way that they appear to come from a point but don't actually travel through this point. In this case, the focal point is *virtual.* You can't see this point on a card, but if you're doing ray tracing, you can trace the rays back along the direction they're traveling and see where they all intersect at the virtual focal point.

If you hold a magnifying glass in front of a distant light source (such as a light bulb), you see an image of the light source on a card placed at the right position behind the magnifying glass. Being able to see the image of the light bulb on a card is a characteristic of a real focal point.

Real focal points

Concave mirrors, mirrors that bow in away from the object, focus incident parallel rays to a real focal point. For each ray that's incident on the mirror surface, the reflection equation (see Chapter 4) applies, and the reflected rays are directed toward a common point, as shown in Figure 5-3. Because the light actually reflects and travels through the point, the focal point is real.

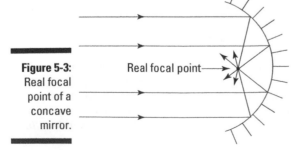

Figure 5-3:
Real focal
point of a
concave
mirror.

Real focal point →

Convex lenses (lenses whose surfaces bow out away from the center of the lens) focus incident parallel rays to a point on the opposite side of the lens (from the incident ray side). Because the lens material is transparent, lenses work with refraction. You apply Snell's law (see Chapter 4) to each ray incident on the surface of the lens and determine the new propagation direction. The net effect of all these refracted rays is that the rays converge to a real focal point, as shown in Figure 5-4. Because the light actually refracts (bends) and travels through the point, the focal point is real.

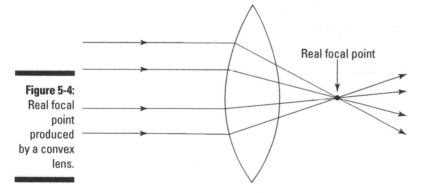

Figure 5-4:
Real focal point produced by a convex lens.

Real focal point

Virtual focal points

Convex mirrors, mirrors that bow out toward the object, diverge incident parallel rays such that they seem to originate from a virtual focal point behind the mirror. For each ray incident on the surface of the mirror surface, you apply the reflection equation, and the reflected rays direct away from a common point, as shown in Figure 5-5. The focal point is virtual because the light actually reflects and travels away from the common point (not through it).

Concave lenses (lenses whose surfaces bow in toward the center of the lens) diverge incident parallel rays such that the rays seem to come from a point on the incident side of the lens, as shown in Figure 5-6. With Snell's law, you can determine the new propagation direction. The rays appear to diverge away from the virtual focal point; you know it's virtual because the light actually refracts (bends) and doesn't travel through the point.

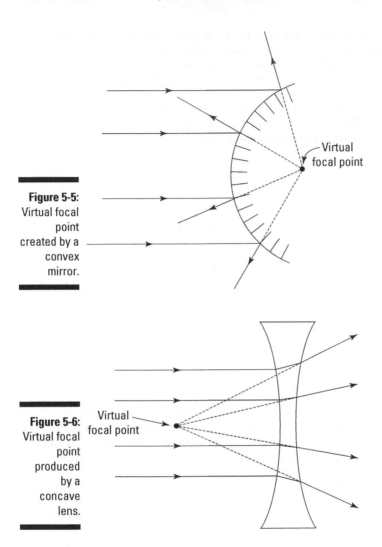

Figure 5-5:
Virtual focal
point
created by a
convex
mirror.

Figure 5-6:
Virtual focal
point
produced
by a
concave
lens.

Chapter 6

Imaging with Mirrors: Bouncing Many Rays Around

. . . .

In This Chapter

▶ Reflecting on flat, convex, and concave mirrors

▶ Examining the characteristics of mirror-produced images

. . . .

Chapter 4 shows you that you can change where light goes by using reflection, where light bounces off a surface. But you can further control the reflection of light with very shiny surfaces called mirrors. *Mirrors* are special light reflectors because they can make images of objects appear at locations other than where the actual object is located. Unlike regular objects, which reflect the incident light in all directions, mirrors reflect light but keep the relative orientation of the incoming rays. This characteristic is why you can see an image in a mirror but not in this page.

In Chapter 4, I show the basic idea for all reflected light using a single ray — that the angle of incidence (the orientation of the incoming light ray) is the same size as the angle of reflection (the orientation of the outgoing, or reflected, ray). But to see the objects around you, you need more than one ray, and using the law of reflection for many rays is a tedious task, at best. Fortunately, you can use geometry to simplify this process so that you have a single, easy-to-use equation.

In this chapter, I show you the equations that make finding the locations and characteristics of images produced by various types of mirrors easy. I expand on the principles covered in this chapter in several chapters that use mirrors to redirect light, but especially in Chapter 19, which looks at how telescopes work with images made by mirrors.

Keeping it Simple with Flat Mirrors

The simplest reflecting surface is a shiny, flat surface. Most practical mirrors are *flat mirrors,* or metal-coated pieces of glass.

The primary purposes of a flat mirror are to present an image in a desired fashion and to redirect a light beam. For example, you want your bathroom mirror to show you what you look like to other people, so you don't want it to present a distorted image. Fortunately, the characteristics of flat mirrors' images are always the same:

✔ The images appear as far behind the mirror as the objects are in front of the mirror.

✔ The images are always virtual, erect, and the same size as the objects.

These characteristics of an image formed by a flat mirror are shown in Figure 6-1.

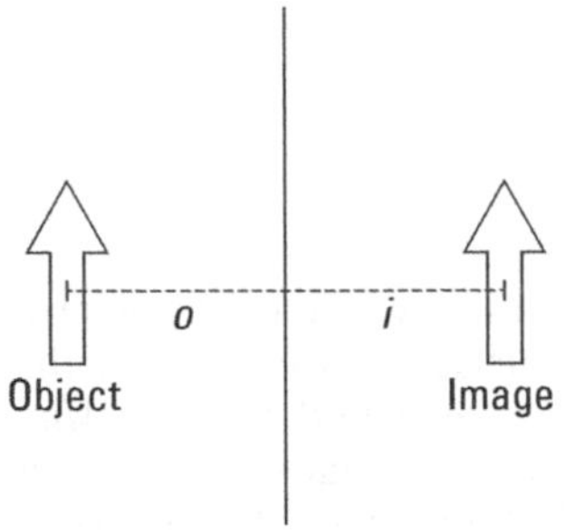

Figure 6-1: Image formed by a flat mirror.

Object   Image

The lateral magnification equation from Chapter 5 is what helps you define the characteristics in the second bullet in the list:

$$M = -\frac{i}{o} = 1$$

In this equation, i is the distance between the image and the mirror (which, in the case of a flat mirror, is a negative number), and o is the distance between the mirror and the object. Because i is always equal to $-o$, the magnification is always equal to 1. This magnification means that the image formed by a flat mirror is the same orientation as the object (erect, unless you're standing on your head), the same size as the object, and virtual, which is why you can't see the image on a screen.

Changing Shape with Concave and Convex Mirrors

Although flat mirrors (see the preceding section) have many uses, they can't do every job. Sometimes you want a mirror to produce an image that doesn't exactly match the original object. To change the characteristics of mirror-reflected images, you have to change the properties of the mirror. The simplest property to change is the shape of the surface.

To keep things simple, I talk only about spherical element mirrors in this section. Instead of being a flat surface, the mirror surface follows a round curve, like a section you'd cut off of a basketball. All curved mirrors aren't spherical element mirrors, but you can tell how a curved mirror will behave after you discover how spherical element mirrors work.

You encounter two types of spherical element mirrors:

- ✔ **Concave:** For this type of mirror, the surface bows in, forming a cave around the object.

- ✔ **Convex:** For this type of mirror, the surface bows out toward the object.

In the following sections, I show you the equations that allow you to calculate the location of images formed by concave and convex mirrors and to determine the characteristics of those images.

Getting a handle on the mirror equation and sign conventions

Because spherical element mirrors are more complicated than flat mirrors, you need the more-complicated mirror equation to determine the location of the image formed. The *mirror equation* relates the curvature of a curved mirror to the image formed by an object placed at a certain distance from the mirror:

$$\frac{1}{o} + \frac{1}{i} = \frac{2}{r}$$

In this equation,

- ✔ o represents the distance that the object lies from the center of the mirror surface.

- ✔ i represents the distance that the image lies from the center of the mirror surface.

- ✔ r represents the radius of curvature of the surface.

Sometimes the maker of a curved mirror gives you the focal length, (f), of the mirror rather than the radius of curvature, as I discuss in Chapter 5. For a spherical element mirror where you know the focal length instead, use this equation to find the radius of curvature:

$$f = \frac{r}{2}$$

To use the mirror equation correctly, you need to use the correct convention for determining the sign (positive or negative) of the number (*o, i, r,* or *f*) to put in the equation. Table 6-1 shows you what the typical sign convention is for mirrors as well as what the sign of the number means.

Table 6-1	Sign Convention for Mirrors	
Situation	*Sign Convention*	*Result*
The object is in front of the mirror.	*o* is positive.	Real object
The object is in back of the mirror.	*o* is negative.	Virtual object
The image is in back of the mirror.	*i* is negative.	Virtual image
The image is in front of the mirror.	*i* is positive.	Real image
The center of curvature is in back of the mirror.	*r* and *f* are negative.	Convex surface
The center of curvature is in front of the mirror.	*r* and *f* are positive.	Concave surface
	M is positive.	Erect image
	M is negative.	Inverted image

After you've got the sign convention straight (or is that curved?), the basic procedure for finding the locations of images produced by these mirrors is as follows:

1. **Determine the front and back of the mirror by finding where the object is located.**

   The side of the mirror that the object is on is the front; the other side is considered the back.

2. **Determine where the center of curvature lies (in front of the mirror or in back).**

3. **Use the mirror equation with the sign convention listed in Table 6-1 to find the location of the image.**

Working with concave mirrors

If you look at the mirror equation with the concave mirror, you may think an infinite number of arrangements is possible, depending on the distance that you choose to set the object from the mirror. However, only three different general arrangements of object locations produce different types of images. Here they are:

✔ If you place an object at a distance greater than a focal length in front of a concave mirror (farther away from the mirror than the focal point is), the image is always real, inverted, and reduced. Your image distance is always a positive number from the mirror equation.

✔ If you place an object at a distance that is smaller than the focal length of a concave mirror (closer to the mirror than the focal point is), the image is always virtual, erect, and enlarged. Your image distance is always negative.

✔ If you place an object at the focal point in front of a concave mirror, the image forms at infinity. *Remember:* Don't confuse the result "the image forms at infinity" with the result "no image." An image that appears at infinity just forms very far away, and you can use a lens to bring it closer. But you don't have a lot of options for recovering an image that doesn't form.

Suppose you have a small toy car placed 20 inches in front of a curved mirror as in Figure 6-2. You know the mirror is concave because it tends to form a cup (or cave) around the car. The center of curvature and the object are on the same side of the mirror, and the mirror has a radius of curvature of 10 inches. To find where the image appears (i), you use the mirror equation and the information in Table 6-1:

$$\frac{1}{o} + \frac{1}{i} = \frac{2}{r}$$

$$\frac{1}{20 \text{ in}} + \frac{1}{i} = \frac{2}{10 \text{ in}}$$

You find that $i = 6.67$ inches.

Many people forget to invert the final number in their calculations. Calculators make solving this equation easy, but you must remember to invert that answer to get the actual image distance. In this example, if you got 0.15 rather than 6.67, you forgot to invert $1/i$.

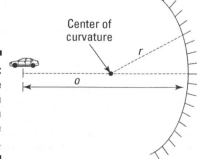

Figure 6-2: Finding the image of a car in a concave mirror.

Center of curvature

r

o

Concave mirrors: Light ray directors

Concave mirrors have the capability to concentrate light and allow you to direct light where you want it. You use concave mirrors when you use your car's headlights when driving at night (or at least I hope you turn your lights on!). Although not perfect spherical elements, headlights are able to capture the light from a light bulb (which sends light out in all directions) and reflect and concentrate the light forward. This capability makes better use of all the light that comes out of the bulb because it directs as much of the light as possible in the useful direction (toward the road) and not in unhelpful directions, such as back into your eyes! An automotive lighting engineer places the light bulb slightly beyond the focal length of the mirror by the proper distance so that the setup casts the light in the desired pattern — wide on the road in front so that you can see where you're going — and as efficiently as possible.

According to Table 6-1, the fact that the image distance is positive means that the image appears in front of the mirror and is a real image. Therefore, you can see the image of the car on a card placed 6.67 inches in front of the mirror (and slightly off center, because a card directly in front of the mirror would block the light from hitting the mirror). To determine the remaining characteristics (magnification and orientation) of the image, you need to calculate the lateral magnification. The equation for that is

$$M = -\frac{i}{o} = -\frac{6.67 \text{ in}}{20.0 \text{ in}} = -0.33$$

The magnification value of –0.33 means that the image is inverted (upside-down) and about a third the size of the actual car.

Exploring convex mirrors

Even though the orientation of the surface relative to the object is different, convex mirror situations use the same mirror equation and sign convention (Table 6-1) as concave mirrors (see the earlier section "Getting a handle on the mirror equation and sign conventions"). If you look at Table 6-1, you notice that for a convex mirror, the center of curvature is in back of the mirror, so you have to treat the r value and the focal length, f, as negative numbers when placing them in the mirror equation to solve for the image location, i.

Like the flat mirror (see the earlier section), convex mirrors always produce virtual images. Because the radius of curvature of a convex lens is always negative, the image distance is always negative. When you calculate the magnification, you always get a positive number less than 1, which means that the image always appears behind the mirror, is erect, and is smaller than the original object.

Suppose you're looking at a large, shiny spherical bowl on the kitchen counter; a small measuring spoon is 8 inches away from the bowl (see Figure 6-3). You want to find where the image will appear and what it will look like. The bowl's surface bows out toward the spoon, so you know it's a convex surface. If the bowl's radius of curvature is 18 inches, you can use the sign convention in Table 6-1 to set up the mirror equation:

$$\frac{1}{o} + \frac{1}{i} = \frac{2}{r}$$

$$\frac{1}{8 \text{ in}} + \frac{1}{i} = \frac{2}{-18 \text{ in}}$$

When you solve this equation for i, you find that it's –4.24 in. Remember that to find the correct answer for the image distance, i, you need to invert your solution to the equation!

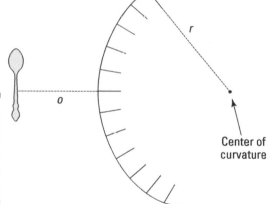

Figure 6-3:
Finding the
image of a
spoon in a
convex
mirror.

Center of
curvature

The negative number means that the image is virtual and that it appears behind the surface of the bowl (you see the image in the surface, not in front of it). You can find the image magnification and orientation by calculating the lateral magnification as follows:

$$M = -\frac{i}{o} = -\frac{-4.24 \text{ in}}{8.0 \text{ in}} = 0.53$$

The magnification value of 0.53 means that the image is erect (same orientation as the spoon on the counter) and about half the size of the actual measuring spoon.

Convex mirrors: The eagle's eye view

Convex mirrors can capture a very wide field of view. Compared to the concave mirror, the convex mirror can bring an object and a large amount of the scenery around the object into an image. This feature is why convex mirrors work well as security monitors in places such as shopping malls or convenience stores and as safety devices in busy buildings that have intersecting hallways.

For example, in a convenience store, the clerk can stay at the cashier's counter but see nearly the entire store in the concave mirror. Unlike a motionless security camera, the mirror allows the clerk to see a different perspective as he moves so that he can concentrate on seeing what a customer (or potential shoplifter) is doing. In a hallway intersection, the convex mirror lets you see whether someone is coming down the hallway, which is especially helpful if the person is walking very quietly. This setup helps avoid accidents, especially when you're pushing a loaded cart to another room.

Chapter 7

Imaging with Refraction: Bending Many Rays at the Same Time

In This Chapter

▶ Finding the locations and characteristics of images formed by one or more refracting surfaces

▶ Discovering the two basic types of lenses and the image locations and characteristics they produce

*Y*ou probably deal with refraction every waking moment, but not with one ray at a time as Chapter 4 describes. *Refraction* is the process where a light ray bends when the ray travels from one material to another. As Chapter 4 shows, Snell's law allows you to calculate how much the refracted ray bends. However, the images you see with your eyes are created by many refracted rays of light.

Dealing with multiple light rays can be tedious both when you have a bad headache and when you're trying to use Snell's law to calculate where an image will appear. Because the images you see often appear in different locations than the objects that cause them, optical systems designers need a tool that allows them to predict where an image will appear so that objects viewed through cameras, telescopes, and so on actually look like users expect them to. Rather than apply Snell's law to every possible ray, they use geometric principles to come up with an equation that is much easier to use.

In this chapter, I show you the equations that make finding the location of images produced by objects in front of a refracting surface easy. I expand on the principles covered in this chapter in Chapters 14 and 21 as you look at how telescopes, microscopes, and projectors work.

Locating the Image Produced by a Refracting Surface

In Chapter 4, I discuss how refraction works with a single ray, but an image requires more than one ray, so you need a model that is more general and much easier to use than Snell's law. Fortunately, some good mathematicians have provided a set of equations that fit this bill. The equations I present in this section allow you to easily calculate where an image will appear relative to a surface.

To begin, you need to identify a situation where refraction (as opposed to reflection) is responsible for creating an image. Consider the factors that cause you to see images of objects rather than the actual objects (flip to Chapter 5 for more):

- ✔ Light bounces off of or is produced by an object and is directed toward your eye so that you can detect it.
- ✔ Between your eye and the object, the medium changes.
- ✔ The light passes through a surface or boundary between two media, causing the light ray to bend at the boundary.
- ✔ The bending of the light rays creates the appearance that the light came from somewhere other than from the actual object. This new point is the location of the image.

Refraction occurs only at the boundary between two materials, so when you're dealing with refracting surfaces, the *surface* is what you call that boundary.

All of these factors come into play when you look at a fish in a fish tank. Finding the location of an image created by the refraction of many rays can be complex, especially if the surface is curved. Fortunately, some relatively simple equations can do this locating for you. The trick is to use the right equation with the particular refracting surface and to follow the proper sign convention, which I show you how to do in the following sections.

Calculating where an image will appear

Snell's law is the equation that tells you how much the path of the light bends as it goes from one material to another. (I cover this law in more detail in Chapter 4.) If you had a lot of time and patience, you could use Snell's law, a protractor, and a ruler to pictorially determine where an image would appear. Luckily, the great mathematical art of geometry provides you with a way to generate a simple equation that accounts for multiple rays refracting at a surface to form an image a certain distance from the surface. For the earlier example of the fish in the fish tank, you can calculate where the image of the fish appears relative to the actual location of the fish.

Your eye-brain system traces the rays back along the line they entered your eye (see Chapter 4). The rays seem to come from the image, which in general isn't where the object is actually located. (This result is the effect of refraction with images: It makes an image at a point that is different from where the actual object is.)

Streamlining Snell's law: Introducing a handy equation

Applying Snell's law to each of several rays allows you to calculate where the image will appear, but geometry allows you to come up with an equation that is much simpler to use.

The equation for the relationship between an object and its image involving a general spherical refracting surface between two different materials is the refracting surface equation

$$\frac{n_1}{o} + \frac{n_2}{i} = \frac{n_2 - n_1}{r}$$

In this equation,

- ✓ o represents the distance that the object lies from the surface.
- ✓ i represents the distance that the image lies from the surface.
- ✓ n_1 represents the index of refraction of the medium that the object is in.
- ✓ n_2 represents the index of refraction of the medium that is on the opposite side of the surface from the object.
- ✓ r represents the radius of curvature of the surface.

Figure 7-1 shows the general arrangement.

Figure 7-1:
General refracting surface arrangement.

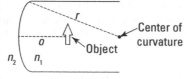

The distinction between front and back is important. The *front* of the surface is the area that contains the object; the *back* is the area on the other side of the surface from the object (this area doesn't contain the object). Being able to identify these two regions is critical to placing the correct sign with the distances in the model.

Assigning the appropriate sign conventions

The next thing you need in order to solve imaging problems at a surface is to understand the *sign convention* — specifically, when to assign a negative or positive sign to a distance (the quantities o, i, and r). This designation is important because the sign of the answer for the location of an image tells you whether the image appears in front of the surface or behind it. For most real-world problems, the index of refraction values (see Chapter 4) are always positive. Table 7-1 summarizes the sign convention used with the equation in the preceding section.

Table 7-1	Sign Convention for Refracting Surfaces	
Situation	**Sign Convention**	**Result**
The object is in front of the surface.	o is positive.	Real object
The object is in back of the surface.	o is negative.	Virtual object
The image is in back of the surface.	i is positive.	Real image
The image is in front of the surface.	i is negative.	Virtual image
The center of curvature is behind the surface.	r is positive.	Convex surface
The center of curvature is in front of the surface.	r is negative.	Concave surface

Head to "Taking a closer look at convex and concave lenses" later in the chapter for more on concave and convex surfaces; Chapter 5 provides more info on real and virtual images.

Solving single-surface imaging problems

You can use the refracting surface equation earlier in the chapter and the information in Table 7-1 to find image locations for almost all refracting surface situations. (Check out the earlier section "Calculating where an image will appear" for this info.) I expand this idea for practical devices (lenses) later in this chapter.

Say that you're looking at a fish that's in a rectangular fish tank. The fish that you see is actually the image of the fish; the actual fish is at a different location. For this example, say that you know where the fish actually is and that you want to calculate where the image will appear. You need to know what value to enter for r, the radius of curvature of the surface, but because the refracting surface is *planar* (flat) as shown in Figure 7-2, the center of curvature is very far away. This fact makes r such a large value that the material's index of refraction is negligible compared to it, so the term that involves r in the refracting surface equation is effectively zero.

Rearranging the equation, you find that for flat surfaces, the image location is determined by the following:

$$i = -\left(\frac{n_2}{n_1}\right)o$$

A general characteristic of planar surfaces is that when a real object is in a higher index material, the virtual image always appears closer to the surface and in front of it. For instance, say that you want to find the location of the image of a fish that is 8 inches from the side of the tank you're looking into as shown in Figure 7-2. n_1 is the index of the material in front of the surface. The fish is in the water, so n_1 is 1.33. You're in air, so the light travels from the fish in the water into the air. So n_2 is 1.00. Using these numbers and the rearranged refracting surface equation, you find that $i = -6.02$ inches. According to Table 7-1, the negative value means that the image is in front of the surface, which is what you see when you view a fish in a rectangular fish tank.

After you've found the location of the image, figuring out the other properties of the image is relatively easy. Aside from the image type (real or virtual, which comes from the sign of the image distance in the problem), the magnification tells you the remaining properties. From Chapter 5, you can use the following magnification equation:

$$M = -\left(\frac{i}{o}\right)$$

For the fish example, you know that $o = 8$ inches and $i = -6.02$ inches. Placing these values in the magnification equation gives you a value of 0.75. The fact that M is positive means that the image is erect, and having a value of less than 1 means that the image appears smaller than the object.

Figure 7-2:
Arrangement for finding the image of a fish in a rectangular fish tank.

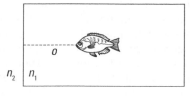

Another common refraction situation involves a curved surface. To keep things simple, I look only at spherical surfaces. It doesn't cover all of the real-world situations, but it does let you see the effect of curved surfaces without the complicated geometry involved in nonspherical curved surfaces.

Say you're looking at a fish in a spherical fish bowl, like the arrangement shown in Figure 7-3. The fish is 5 inches away from the surface closest to you, and the radius of curvature of the surface is 12 inches. Because the surfaces of most spherical fish bowls typically bow out toward you, the radius of curvature is negative; the center of curvature (see Chapter 5) is (in front of the surface) on the same side of the surface as the fish (the object). Using Table 7-1 and the refracting surface equation with $n_1 = 1.33$ for the water and $n_2 = 1.00$ for the air, you find that $i = -4.2$ inches. Like the fish in the rectangular tank, the image is in the water and closer to the surface than the object. Calculating the magnification, you find that $M = 1.12$. The image is erect and larger than the actual fish.

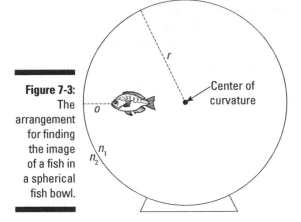

Figure 7-3:
The arrangement for finding the image of a fish in a spherical fish bowl.

Note: This example doesn't account for the actual bowl. For simplicity's sake, I assume that the material has the same index of refraction as the water. In real life, most bowls are made of glass, which has a different index of refraction (1.50) than water (1.33). I talk about how to handle the situation when the bowl has a different index than water in the following section, but in most situations, you can neglect the thickness of the bowl because the difference between the image distance you get by assuming that the tank and the water have the same index and the distance you get by including the refraction at each boundary of the tank glass fish is 2 percent or less. In other words, as long as the fish tank walls aren't too thick (most fish tanks have walls that are about ⅛-inch thick), you don't have to worry about the fish tank.

Working with more than one refracting surface

So what do you do if you have two or more refracting surfaces in a situation? Fortunately, the procedure is no more complicated than dealing with a single surface (see the preceding section). The system is now all the refracting surfaces in the situation that you're trying to deal with; the trick is to determine the front of the system (the side of the system that the object is on). This side represents the front for all of the surfaces in the problem. You treat each surface individually and start with the surface closest to the object (not the observer). The image from this surface then becomes the object for the next surface, but you must change the value of *i* from the first surface to *o* relative to the second surface. You continue this process until you reach the last surface. The image location you find here is the image location for all the surfaces in the system. This method is the basis by which lenses (the workhorses of optical systems) and other interesting natural refracting processes work.

The basic procedure for finding the location of images from refracting surfaces is as follows:

1. **Determine the front and back of the surface closest to the object.**

2. **Figure out how many surfaces are involved.**

3. **Find the surface closest to the object and use the appropriate equation (based on the curvature of the surface) to find the image location.**

4. **Transform the image distance from Step 3 into an object distance relative to the next surface and repeat the calculation in Step 3 to find the next image location.**

5. **Repeat Step 4 until you've dealt with all of the surfaces.**

    The image location found with the last surface is the location of the image that appears through all the refracting surfaces.

Suppose you want to calculate where an image of a fish in a fish bowl will appear when the walls of the fish bowl aren't thin enough to ignore. Because the index of refraction of glass is different than water, you have two refracting surfaces to deal with. To make this example more general, assume you're at an aquarium where the glass is bowed out and thick. The fish in this situation is 24 inches from the inside wall, which is 3 inches thick, and the radius of curvature of both refracting surfaces is 48 inches.

Images formed by refracting surfaces: Mirages

Mirages are the fuzzy images of what appears to be water on the ground or a reflection of trees off water on the ground. As you get closer to where you thought the water or trees were, the image seems to disappear or move farther way. The appearance of these images is a refraction process in the air that results from several layers of air with slightly different indices of refraction. You can easily understand why these images form by repeating the procedure for finding the image locations with more than one refracting surface (check out the nearby section "Working with more than one refracting surface" for more).

Mirages occur on sunny days where the ground temperature is higher than the air temperature due to heating from the sun. The hotter air closest to the ground is less dense and has a lower index of refraction than the increasingly cooler air farther above the ground. This situation creates many refracting surfaces, which has the net effect of constantly bending the light away from the surface normal as the light travels toward the ground.

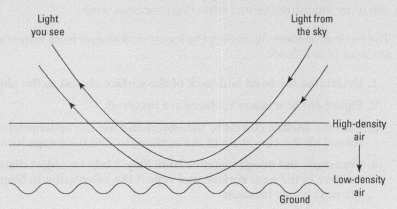

Refraction bends the light from low in the sky into your eye before the light can hit the ground. The image you see is that of the sky, which is blue on a clear day. But your mind is trained to think that blue on the ground is water, so you see the mirage of water on the ground ahead of you. After you get closer to the area, however, the water disappears and moves farther away. You can never get to it. Kind of a mean trick of nature for someone lost in the desert, but fun to look at while you're driving in your air-conditioned car along a long, flat highway.

To calculate where the image appears, you need to start with the surface closest to the fish. Using the procedure for refracting surfaces, you have for the first surface

$$\frac{n_1}{o_1} + \frac{n_2}{i_1} = \frac{n_2 - n_1}{r_1}$$

$$\frac{1.33}{24 \text{ in}} + \frac{1.50}{i_1} = \frac{1.50 - 1.33}{-48 \text{ in}}$$

If you perform your calculations correctly, you find i_1 = –25.44 inches.

To calculate the effect of the second refracting surface (between the glass and the air), you need to transform the image distance from the first surface into a distance measured relative to the second surface. Because the wall is 3 inches thick, the image produced by the first surface is located 28.44 inches (25.44 + 3) from the second surface.

Repeating the math,

$$\frac{1.50}{28.44 \text{ in}} + \frac{1.00}{i_2} = \frac{1.00 - 1.50}{-48 \text{ in}}$$

i_2 = –23.63 inches. This location is where the image appears when you're looking into the aquarium tank relative to the surface of the glass nearest to you. Notice again that the image is still in front of the surfaces but closer to the inner one.

Looking at Lenses: Two Refracting Surfaces Stuck Close Together

Refracting surfaces are neat to look at (see the earlier section "Working with more than one refracting surface"), but if you want to make eyeglasses, for example, you're probably better off to use something other than swimming pools or spherical fish bowls. So lenses are practical, compact devices (that aren't prone to being wet) that allow you to make images appear where you want them instead of letting a surface of water dictate their location.

You can treat most lenses as thin lenses as long as the thickest part of the lens is much smaller than the object or image distances of the lens (which is the case with all general lens problems). This fact simply means that you don't have to account for the separation between the two refracting surfaces to determine the final image location, which greatly simplifies the modeling of complex lens systems. Thick lenses do appear in the real world, but dealing with situations where you consider the thickness of the lens is beyond the scope of this book.

Designing a lens: The lens maker's formula

Because most lenses show up in optical instruments, such as cameras, telescopes, and so on, you need a model so that you can make lenses that form an image where you want it. In Chapter 5, I show you that you can characterize the effect of a mirror or a lens on light rays if you know its focal length. With lenses, knowing the focal length lets you calculate where the image of an object placed in front of the lens will appear. When designing an optical system, this approach is much easier than applying Snell's law over and over again and simpler than the procedure I discuss earlier in the chapter for a thick spherical aquarium.

The model for creating a lens with the desired focal length is an equation called the *lens maker's formula for thin lenses.* This model is much easier to work with than the refracting surfaces equation earlier in the chapter and allows you to find all the important data for making the correct lens.

Two quantities determine the focal point of a thin lens:

> ✔ **The curvature of the two surfaces of the lens.**
>
> ✔ **The change in index of refraction.** This quantity is the difference between the index of refraction of the material that the lens is in (such as air or water) and the index of refraction of the lens material.

These two features are included in the lens maker's formula for thin lenses:

$$\frac{1}{f} = \frac{(n_2 - n_1)}{n_1}\left(\frac{1}{r_1} - \frac{1}{r_2}\right)$$

In this equation:

> ✔ n_1 is the index of refraction of the material that the object is in (air for most situations).
>
> ✔ n_2 is the index of refraction of the material that the light is traveling into (the material the lens is made of).
>
> ✔ r_1 is the radius of curvature of the surface that the light hits first.
>
> ✔ r_2 is the radius of curvature of the surface the light hits last.

Suppose you want to find the focal length of a lens that bows out from its center on both sides (a *biconvex lens*) such as the one shown in Figure 7-4. In order to calculate this length correctly, you need to determine the front side of the lens so that you make sure you keep your signs right. For the sake of this example, I call the left area the front and the area to the right of the lens the back.

Using the sign convention shown in Table 7-1 and starting from the left, notice that the center of curvature of the first surface the light encounters is behind the lens. So if the radius of curvature is 25 centimeters, you enter positive 25 centimeters for r_1 in the equation. The radius of curvature of the second surface is a negative 30 centimeters because the center of curvature for this surface lies in front of the lens. Assuming that the lens is in air (making

n_1 = 1.00) and that the lens is made out of regular glass (making n_2= 1.50), you find that the focal length for this lens is about 27.3 centimeters.

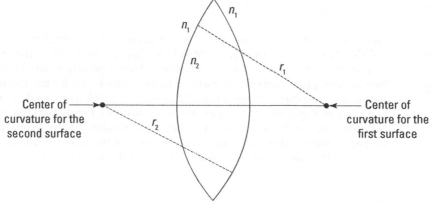

Figure 7-4: Geometry for a biconvex lens.

With that information, you can calculate where the image will appear when you use the lens to create an image of an object. The equation that relates the image location to the object distance and focal length is the thin lens equation:

$$\frac{1}{o} + \frac{1}{i} = \frac{1}{f}$$

Because the thin lens is simpler than the general refracting surface, you can use a less-complicated table for the sign convention. Table 7-2 shows this information.

Table 7-2   Sign Convention for Thin Lenses

Situation	Sign Convention	Result
The object is in front of the lens.	o is positive.	Real object
The object is behind the lens.	o is negative.	Virtual object
The image is behind the lens.	i is positive.	Real image
The image is in front of the lens.	i is negative.	Virtual image
The lens is a converging or positive convex lens.	f is positive.	
The lens is a diverging or negative concave lens.	f is negative.	

Sometimes a lens is planar on one side and concave or convex on the other. This scenario doesn't change the behavior of the lens. As long as one side is convex or concave and the other side is flat, you have a plano-convex or plano-concave lens, and you can still use Table 7-2 to determine the sign of the focal length.

Many other types of lenses, such as meniscus lenses or graded index lenses (GRIN lenses), that I don't present in this chapter don't follow the convention of simply convex or concave. If you have a lens type other than bi-convex, bi-concave, plano-convex, or plano-concave (which are, by far, the most common), you can still use Table 7-2 as long as you know whether the lens is a converging or diverging lens. If you're not specifically told that the lens is converging or diverging, you can determine which it is by using the lens maker's formula. Use this characteristic and not the convex or concave nature of the surface to determine the sign of the focal length.

Taking a closer look at convex and concave lenses

With all the possible values for the variables in the lens maker's formula, you can find basically two types of lenses: convex and concave. A *convex* surface or lens is one whose surface curves out toward the object, making the center of the lens thicker than the outer edges. A *concave* surface or lens is one whose surface curves away from the object (tending to form a cave around the object), making the center of the lens thinner than the outer edges. Regardless of the specific focal length, each lens type presents certain relationships that always exist between the object and its image. The following sections present the general characteristics of the two basic lenses that serve as the basis for lens systems in later parts of this book (so you may want to dog-ear this discussion for easy reference).

Finding the image locations of convex lenses

Convex lenses (also known as *positive* or *converging* lenses based on what they do with light) always have a positive focal length by convention. Using this feature with the thin lens equation, you see an infinite number of possible arrangements. However, only three different general arrangements generate different types of images. These arrangements and the image characteristics are as follows:

- If you place an object at a point greater than a focal length in front of a convex lens, the image is always real, inverted, and reduced. Your image distance is always positive if you've done the calculations correctly.

- If you place an object at a point smaller than a focal length in front of a convex lens, the image is always virtual, erect, and enlarged. If your calculations are sound, your image distance is always negative.

> ✔ If you place an object at a distance equal to the focal length from a convex lens, the image forms at infinity.
>
> Don't confuse the result "the image forms at infinity" with the result "no image." An image that appears at infinity just forms very far away. You can manipulate where infinity appears with an appropriate lens to access the image, but you can't recover an image that doesn't form.

Locating the images of concave lenses

Concave lenses (also called *negative* or *diverging* lenses based on what they do with light) have negative focal lengths by convention. As with convex lenses (see the preceding section), you have an infinite number of possible arrangements when you consider this feature with the thin lens equation. However, unlike the convex lens, concave lenses always produce a virtual, erect, and magnified image. The image distance in the imaging equation is always a negative number if you do the calculations correctly.

Finding the image location and characteristics for multiple lenses

If you have a problem that has two or more lenses in it, the procedure is similar to multiple refracting surfaces (flip to "Working with more than one refracting surface" earlier in the chapter). You use the imaging equation in that section to find the image, but you have to adapt it to become the object for the next lens in line.

In a nutshell, the procedure to find the final image produced by a lens system with two or more lenses is as follows:

1. **Determine the front and back of the system.**

2. **Figure out how many lenses are involved.**

3. **Start with the lens closest to the object and use the thin lens equation to determine the image location.**

   This image becomes the object for the next lens.

4. **Transform the distance between the image and the first lens to a distance measured relative to the next lens.**

   This transformed distance may be a negative number if the image lies behind the second lens. After you've made the proper transformation, you use the thin lens equation to find the image location.

5. **Repeat Step 4 until you've dealt with all the lenses in the problem.**

   The image location you find with the last surface is the location of the last image. The product of all of the magnifications produced by each lens in the system determines the final image orientation and magnification.

Figure 7-5 shows a typical arrangement of two lenses that I work through to finish out the section by showing you how this process works.

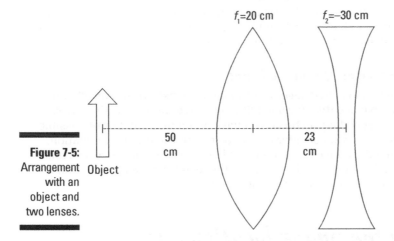

f_1=20 cm f_2=−30 cm

50 cm   23 cm

Figure 7-5: Arrangement with an object and two lenses.

Object

Say that the magnitudes of the focal lengths are 20 centimeters for the first lens and 30 centimeters for the second. Notice the second lens has bowed-in sides; that indicates it's a concave lens, so its focal length is actually −30 centimeters when you use the thin lens equation.

Start by identifying the front of the system. In Figure 7-5, the object is 50 centimeters to the left of the first lens, so the front of both lenses is the area to the left of the lenses.

Plug the values for the object distance and focal length for the first lens in to the thin lens equation to find where the image appears.

$$\frac{1}{o_1} + \frac{1}{i_1} = \frac{1}{f_1}$$

$$\frac{1}{50 \text{ cm}} + \frac{1}{i_1} = \frac{1}{20 \text{ cm}}$$

Solving for i_1, the image distance for the first lens, you get 33.3 centimeters.

Many people forget to invert the final number in their calculations. Calculators make solving for $1/i$ in the thin lens equation easy, but you need to invert that answer to get the actual image distance, which is i. For example, if you got 0.03 rather than 33.3, you found $1/i_1$ and need to invert that number.

Now you need to deal with the second lens. The image created by the first lens becomes the object for the next lens. With the image distance for the first lens (33.3 centimeters), you need to find where this image lies relative to the second lens. In Figure 7-5, an image located 33.3 centimeters from the first lens is actually to the right of the second lens.

To find where the image forms with the second lens, you must find the distance from the second lens to the first image. In this case, the two lenses are separated by 23 centimeters, so the image from the first lens appears 33.3 centimeters – 23 centimeters = 10.3 centimeters to the right of the second lens. The 10.3 centimeters becomes the object distance for the second lens. According to Table 7-2, when the object is behind the lens, the object distance is negative. So, using the thin lens equation

$$\frac{1}{o_2} + \frac{1}{i_2} = \frac{1}{f_2}$$

$$\frac{1}{-10.3 \text{ cm}} + \frac{1}{i_2} = \frac{1}{-30 \text{ cm}}$$

Solving for i_2, the image distance for the second lens, you get 15.8 centimeters. Table 7-2 tells you that because the image distance is a positive number, the image formed by the two-lens system is a real image.

In most cases, you need to determine the other characteristics of the image, specifically the relative size of the image and the orientation, which I discuss in Chapter 5. You determine both by calculating the lateral magnification of the image. The magnification of the two-lens system is the product of the magnifications produced by each lens individually. So, for the current situation,

$$M_1 = -\left(\frac{i_1}{o_1}\right) = -\left(\frac{33.3 \text{ cm}}{50.0 \text{ cm}}\right) = -0.667$$

$$M_2 = -\left(\frac{i_2}{o_2}\right) = -\left(\frac{15.7 \text{ cm}}{-10.3 \text{ cm}}\right) = 1.53$$

$$M = M_1 \cdot M_2 = (-0.666) \cdot (1.52) = -1.02$$

The final image produced by the two-lens system in Figure 7-5 appears 15.7 centimeters to the right of the second lens, appears slightly larger than the original object, and is inverted (upside-down with respect to the original object). If you have a system with more lenses, you just repeat the image-object calculations until you've worked with all of the lenses in the problem.

D'oh, fuzzy again! Aberrations

As you may be aware from operating your camera in the manual focus mode, you can't just place a lens anywhere and get a clear image where you want it to appear. For one thing, you have to move the lens until the image is clear, or in focus, to keep from getting a blurry image. And if your lens is shoddy, forget about image clarity. But even when all conditions are ideal, *aberrations* prevent the image from being clear. The two main aberration types (spherical and chromatic) are nothing more than the physical consequences of using spherical element lenses and refraction.

✔ **Spherical aberration:** *Spherical aberration* isn't a manufacturing defect but rather a simple consequence of a spherical surface. Rays of light that hit the optic at the outer edge of the optic are focused to a different point than *paraxial* rays (rays that hit the optic near the center). This discrepancy results in the rays not concentrating at one spot, as shown in Figure 7-6. Because the image has many focal points, it appears blurry regardless of where the lens is located (although changing the location may affect which parts of the image are blurry).

You can reduce or eliminate spherical aberration by using an *aperture* (a piece of paper or sheet of metal with a hole in it) to block the outermost rays from hitting the outer edges of the optics. Although this device can make the image clearer, using this technique reduces the image's brightness. Photographers have to balance these two factors when taking pictures.

✔ **Chromatic aberration:** *Chromatic aberration* happens only with refraction optics, such as lenses. Because of dispersion (see Chapter 4), the index of refraction varies by wavelength, so different-colored rays focus to different distances from the lens as shown in Figure 7-7. Because you have multiple focal points, the image appears blurry and the objects appear to have a rainbow outline.

Chromatic aberration doesn't happen with mirrors because they use reflection, which doesn't depend on the index of refraction (whereas lenses depend on refraction, which includes dispersion). That's one reason why astronomical telescopes are typically made with mirrors rather than lenses to try to reduce or eliminate chromatic aberration.

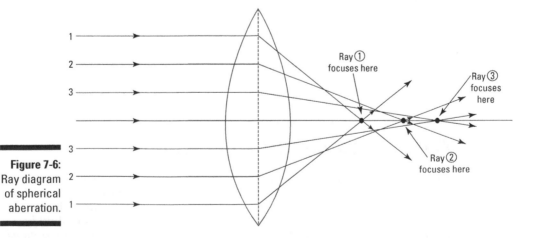

Figure 7-6:
Ray diagram of spherical aberration.

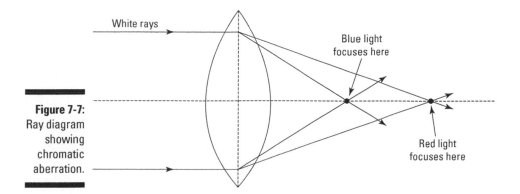

Figure 7-7:
Ray diagram
showing
chromatic
aberration.

Part III

Physical Optics: Using the Light Wave

The 5th Wave — By Rich Tennant

"As you can see, the Aurora Borealis is quite strong this far north."

In this part . . .

Part III concentrates on the wave property of light. I discuss properties associated with waves, such as the orientation of the electric field and optical devices that can change this orientation. Like other wave phenomena, light waves can interact with each other (called *interference*) if the conditions are just right. Optical interference is a very subtle experience, but if you blow bubbles in the air, you can see interference in the colors of the bubbles.

By understanding optical interference and how to create it, you can arrange different set-ups to measure the optical properties of many different materials and the physical properties (such as how smooth the surface is) of manufactured items. This part also explores another wave property of light — its ability to bend around obstacles — that creates interesting patterns leading to important optical devices for determining what specific types of materials are present.

Chapter 8

Optical Polarization: Describing the Wiggling Electric Field in Light

. . . .

In This Chapter

▶ Getting familiar with optical polarization

▶ Discovering polarization types

▶ Polarizing light

. . . .

*O*ne way to look at light is as a transverse electromagnetic wave (see Chapter 3 for details). The orientation of the transverse stuff in this wave picture is important when you're looking at light bouncing off a surface (reflection) or passing through materials (refraction). The effects often depend on how the stuff wiggling in light waves — the electric and magnetic fields — align with the structure of the material.

In this chapter, I define optical polarization and explain why it's defined in relationship to the light waves' electric field (versus their magnetic field). I describe the different types of polarization and how you can make polarized light.

Describing Optical Polarization

Two models exist for light: You can view light as a particle or as a wave. In the wave picture, the thing that's waving is actually two items — an electric field and a magnetic field (see Chapter 3 to get a better picture of this arrangement).

When you're trying to understand why light works with a material the way it does, you need to know how the wave affects the material that it's hitting or passing through; quantitatively, you need an appropriate way to talk about the orientation of the electric field in the light wave as it interacts with atoms or molecules in the material. The field of optics uses the term *optical polarization* (often shortened to *polarization*), which describes the orientation of the electric field in the light waves. Many optical scientists and electrical engineers prefer to use the formal (longer) name to distinguish optical polarization from *dielectric polarization,* which, for example, is what happens in dielectric materials (electrically insulating materials) inside capacitors. The distinction on this particular point needs to be clear because light can cause dielectric polarization (see Chapter 21).

Because electric fields affect charged particles, the electric field in light waves can affect the dielectric polarization of a material. *Dielectric polarization* is the tendency of electric fields to create dipoles or cause randomly oriented dipoles to line up parallel to each other. This phenomenon is important in capacitors because it reduces the electric field between the plates of the capacitor so that more charge can be placed in the capacitor for a fixed potential difference (in other words, you can store more energy in the capacitor). Clarifying what you mean when you say *polarization* early is important when working on multidisciplinary teams so that everyone understands what polarization you're talking about.

Focusing on the electric field's alignment

Electromagnetic waves contain two field quantities: electric and magnetic. These fields are always perpendicular to each other and to the direction that the wave is moving. If you know the orientation of the electric field and the direction of the wave, you know the orientation of the magnetic field because the magnetic field is perpendicular to the electric field and the direction that the wave is traveling. When you develop a mathematical model, you'd probably prefer to have to worry about only one field, but which one do you choose?

Optical scientists choose the orientation of the electric field as the one to watch. The whole reason for trying to describe the orientation of the wiggling fields is to understand what happens when light interacts with a material. Because the materials that you work with are made of atoms, which contain electrons and protons, you should choose the field that will significantly affect charged particles.

According to the fundamental force laws of magnetism and electricity, the electric field is able to do mechanical work on charged particles. This ability means that an electric field causes the charged particle to move (actually, to accelerate). The magnetic field force appears only when the charged particle is moving; the magnitude of the force depends on how fast the particle is moving. On top of that, the magnetic field force is always a *centripetal force,* meaning it causes the particle to start moving in a circle. Because a centripetal force is always directed toward the center of the circular path, the magnetic field force can't change the speed of the particle and therefore can't do any work on a charged particle. Figure 8-1 shows the differences between the two forces.

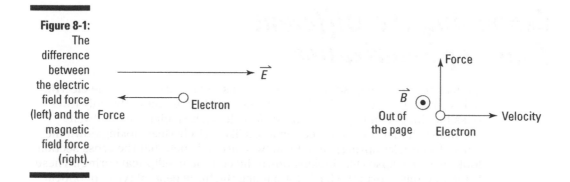

Figure 8-1:
The difference between the electric field force (left) and the magnetic field force (right).

For you to detect the effect of light interacting with the atoms in a material, a force must do work on the atoms in the material. Magnetic forces are negligible in most cases because they can't do work on the charged particles in the materials, so you typically neglect them in optics. When you study light and matter interactions, the electric field is what does all the work. The magnetic field is still present; you just see only the effect of the electric field when studying optics.

Polarization: Looking at the plane of the electric field

As I mention in Chapter 3, a sine function represents the electric field. But this sine function tells you how the magnitude of the field changes in space and time only. The electric field is a vector, so it has a direction as well. Although the sine function adequately describes the way the magnitude of the field changes, the direction requires a different equation.

Optical polarization is a way to describe the surface that contains the oscillating electric field vector in electromagnetic waves. For light waves, this surface can be a plane like a wall, which never changes. The surface can also be like a slightly twisted piece of paper. If the electric field vector points parallel to the surface, notice that the electric field vector can twist around an axis that is parallel to the propagation direction of the wave. Regardless of what kind of surface you have, note that the direction of the electric field doesn't change randomly but rather in a smooth and predictable fashion. This characteristic of the electric field vector in electromagnetic waves is the only reason we can talk about different polarization states.

Examining the Different Types of Polarization

Optical polarization describes what happens to the surface that the electric fields wiggle in. Because the direction doesn't change randomly, only three possible things can happen to the surface: It can be a plane, it can twist around such that the cross section looks like a circle (like looking down the coils of a circular spring), and it can twist around such that the cross section looks like an ellipse (like looking down the coils of an elliptical spring). These three outcomes form the basis for naming the three general types of optical polarization: linear, circular, and elliptical. In the figures in this chapter that show the different polarization states, I present what the light looks like as it travels toward you. The sinusoidal motion still exists and is usually provided in most textbooks, but I show you the end result, which provides the condition for naming the different polarization states of light.

The bottom line is that polarization, whether light is polarized or unpolarized, is often a major factor that must be considered in the construction of many optical systems. The following sections take a look at the various kinds of light polarization and how they can play into real-world applications.

Remember that I'm showing only what happens to the orientation of the electric field, because the magnetic field part of the light wave has a negligible effect on charges.

Linear, circular, or elliptical: Following the vector path

Optical polarization can describe a single wave or a collection of many electromagnetic waves. When talking about many waves, you can consider the entire group polarized if all the individual waves have the same polarization state and orientation. The behavior of a large number of waves is very important when designing optical devices, such as fiber-optic communication links, that send information. When modeling such systems, analyzing the effect on one wave and generalizing for many waves is often convenient. In this section, I describe the polarization states that can be applied to both a single and a collection of light waves. The later section "Random or unpolarized: Looking at changing or mixed states" presents additional polarization types that apply only to a collection of waves.

Linear polarization

The simplest polarization state to understand is called *linear polarization*. Some people call it *plane polarization* because the electric field oscillates in a surface that is planar, but I prefer linear polarization because when you model the effect of a material on this type of polarized light, it seems easier to think of it in terms of a line.

Linearly polarized light basically means that the electric field oscillates in a plane. The left side of Figure 8-2 gives you an idea of what this oscillation looks like. The figure shows that the electric field magnitude varies in time sinusoidally, but the direction remains within the plane of the page (although it changes from pointing up to pointing down).

Figure 8-2:
Linearly
polarized
light from
the side
(left) and
from
straight on
(right).

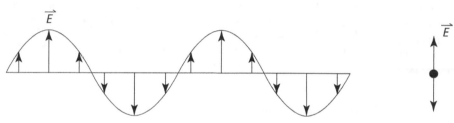

If you aren't dealing with an experiment in quantum optics, you have a gazillion more light waves to deal with. What does that mean for polarization? For you to detect linearly polarized light, a vast majority of the light waves must have the same orientation, and the planes must be parallel to each other. Figure 8-3 shows what a large number of linearly polarized light waves traveling perpendicular to the page must look like for you to be able to call the light you see linearly polarized.

Figure 8-3:
Linearly
polarized
light for a
large collec-
tion of light
waves.

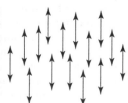

Light can have any orientation as long as the electric field remains in a plane and the planes of all the light waves present are parallel to each other. The linearly polarized light that I have used in this section has its electric fields oscillating up and down. You can think of this state if you hold a piece of paper so that it's vertical; the electric field oscillates parallel to the paper. This orientation is called *vertically polarized light.* Another common type of linearly polarized light occurs when the plane is oriented horizontally and is known as *horizontally polarized light.* (To visualize this state, lay the piece of paper down on a table; the electric field oscillates parallel to the paper.) Figure 8-4 shows vertical, horizontal, and other types (ones without special names) of linearly polarized light with the light traveling perpendicular to the page.

Figure 8-4:
Various types of linearly polarized light.

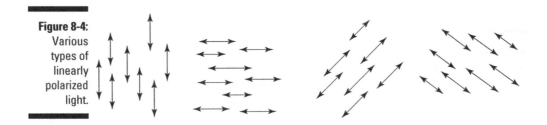

Circular polarization

Vector quantities are nice when they lie along one axis or the other, such as the vertical axis or the horizontal axis. However, when they lie at some angle relative to the reference axes, you need to split the vector into components and look at what happens to each component separately. Circular polarization describes the case where the surface that the electric field oscillates in twists around an axis parallel to the direction that the wave travels. You can view the tip of the electric field vector as following the coils of a spring; check out the nearby sidebar for more on using a real spring as a 3-D model.

To model circular polarization (especially when you want to see what happens to the light traveling in a birefringent material), the instantaneous electric field vector is split into two orthogonal components; vertical and horizontal are usually convenient. (See the later section "Birefringence: Splitting in two" for more on birefringence.) Each component has the same amplitude E_0 and varies in time sinusoidally such that when you compute the resultant, the tip of the resultant electric field vector traces out a circle as time passes. Each component's magnitude follows a sine function, but one is shifted by 90 degrees relative to the other, which causes it to follow a cosine function. The result of the phase difference between the two components is that when one component goes through a maximum value, the other one is zero.

Visualizing circular polarization with a spring

You can use a simple spring, like one from a retractable ballpoint pen, to get an idea what circular polarization looks like because the coils of the spring give you a good physical model for the path traced out by the electric field of circularly polarized light. If you hold the spring horizontally and look at it from the end, the spring looks like a circle (hence the name circular polarization). This view is the path traced out by the electric field of light traveling straight toward you. If you rotate the spring and look at it from the side, the spring resembles a sine wave, which is what the vertical component of the electric field follows. If you look down on the spring from above without rotating it from its horizontal position, you see the same sinusoidal form, but slightly shifted relative to the view from the side. This view gives you an idea about the horizontal component, emphasizing the phase shift between the two components.

This 90-degree difference between the two orthogonal components causes the *resultant electric field vector* (which is the vector sum of the two components at each instant of time) to swing around the origin, tracing out a circle. Figure 8-5 illustrates this swing for a few orientations. The left side of the figure shows the two components starting with position 1 (where the vertical component E_y is at a maximum and the horizontal component E_x is zero), moving to position 2 where the two components have the same magnitude of $0.7\,E_0$, then position 3 (where the vertical component is zero and the horizontal component is at a maximum), and so on. Notice that the resultant electric field vector always has the same magnitude. The direction is what changes smoothly with time to sweep out a circle.

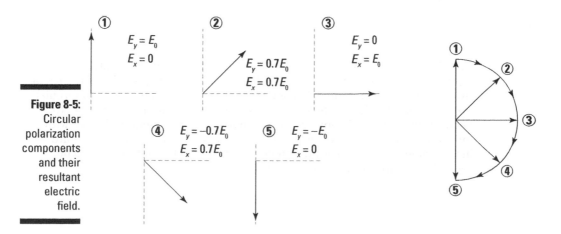

Figure 8-5:
Circular polarization components and their resultant electric field.

The two components of the electric field vector have a phase difference of 90 degrees. But the resultant electric field of the two components rotates one way or the other (clockwise or counterclockwise), depending on which component leads the other. This difference creates two types of circular polarization, *left circularly polarized light* and *right circularly polarized light*. In most of the situations that you encounter, which direction the electric field rotates doesn't matter; the important point is only that the light is circularly polarized.

Different works use different conventions for circular polarization. The left- or right-handed rotation is dependent on the author; you don't find a universal standard for left or right circular polarization. Because I'm the author of this book and I like to take things head on, I classify circularly polarized light based on me looking at the light coming directly toward me. This orientation appears in Figure 8-6 with the resultant electric field rotating in the direction indicated by the arrow. Other authors, including ones much more notable than I, choose to classify circularly polarized light based on what the light waves would look like traveling away from them, not toward them. One way isn't more correct than another.

Figure 8-6:
Left and right circularly polarized light shown as it travels directly toward you.

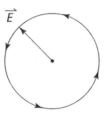
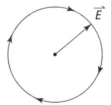

Just like with linearly polarized light, you usually deal with a huge number of light waves rather than just one. For linearly polarized light, all the planes of oscillation for the electric fields need to be parallel to each other. For circular polarization, the electric field vectors rotate, so they aren't all parallel to each other. Figure 8-7 shows what circularly polarized light looks like with multiple light waves. The electric fields aren't parallel, but they all rotate in the same direction, a condition necessary for all the light waves to be regarded as circularly polarized light.

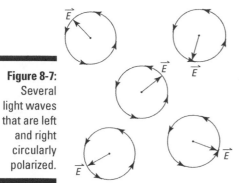
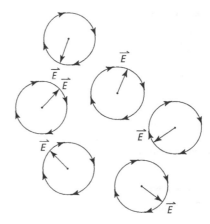

Figure 8-7:
Several
light waves
that are left
and right
circularly
polarized.

Elliptical polarization

The most common type of polarization is *elliptical polarization,* in which
the electric field vectors rotate in a similar fashion to those for circular
polarization (see the preceding section).

For circular polarization, the vertical and horizontal components have the
same *amplitude* (maximum value), and the phase difference between the two
components' oscillation is exactly 90 degrees. However, you get elliptical
polarization instead if the vector components have different amplitudes
or if the phase difference isn't 90 degrees:

- ✔ **Different amplitudes:** If the components have different amplitudes but
  the phase difference is still 90 degrees, the resultant electric field traces
  out an ellipse rather than a circle because one component is greater
  than the other one. With a phase difference of 90 degrees, the major and
  minor axes of the ellipse line up with the vertical and horizontal axes.
  Figure 8-8 shows two cases of elliptical polarization. The figure on the
  right shows what happens if the horizontal amplitude is larger than the
  vertical, and the figure on the left shows what happens when the
  vertical amplitude is larger than the horizontal.

- ✔ **Phase difference other than 90 degrees:** Another way to generate
  elliptically polarized light is to have components with the same amplitude
  but a phase difference between them of something other than 90 degrees.
  When the phase difference isn't 90 degrees, one component won't zero
  when the other component reaches its maximum value (the requirement
  for circular polarization). For this kind of elliptical polarization, the
  major and minor axes don't line up with the horizontal and vertical
  axes, as shown in Figure 8-9. This figure demonstrates an elliptical polarization with a phase difference of 45 degrees and equal amplitudes for
  the two components.

Figure 8-8:
Elliptical
polarization
states for
compo-
nents with
different
amplitudes.

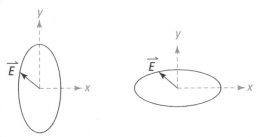

Figure 8-9:
Elliptical
polarization
for a phase
difference
of
45 degrees.

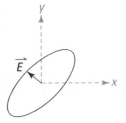

All elliptical polarization states have a rotation associated with them in the same manner as circular polarization. The direction of the rotation is determined by which component leads the other. For most of the situations that you have to deal with, except for maybe homework problems, the direction of the rotation doesn't matter. To classify a group of many light waves as elliptically polarized, you don't need the electric field vectors to be parallel, but the major and minor axes of the paths the electric field vectors trace out must be parallel to each other, as Figure 8-10 indicates.

Figure 8-10:
Left and
right
elliptically
polarized
light for
several light
waves.

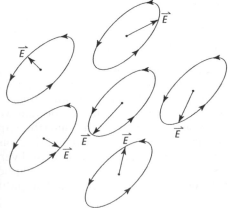
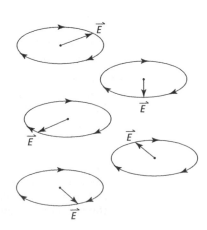

As with circular polarization, the convention for labeling elliptical polarization left or right rotation depends on the author. The convention I use is to look at the rotation of the wave's electric field as the light is traveling toward me. Using this convention, Figure 8-10 shows elliptical polarization states for a few light waves for both left and right rotation.

Random or unpolarized: Looking at changing or mixed states

When working with practical applications of light, especially with fiber-optic systems, you're concerned with the behavior of many light waves, not just one. Looking at a large number of light waves in terms of polarization introduces a couple of polarization states, random polarization and unpolarized light, that relate specifically to a collection of waves and not to single waves.

Random polarization

Random polarization is a term used to describe what can happen when polarized light is used with fiber optics. Many laser sources produce linearly polarized light. When you put this polarized laser light into a fiber, the polarization state often gets distorted as the light bounces around inside the fiber. Most often, the polarization state gets converted to an elliptical polarization state, but the orientation of the major and minor axes are random.

This randomness usually occurs because of bends that you put in the fiber when you set up the experiment. The light that comes out the output end of the fiber is still polarized; you just don't know exactly what polarization state you have unless you measure it. If you change anything, even slightly, the polarization changes randomly to some other state. I've been frustrated by this phenomenon in my lab as changes in temperature (especially in the winter) convert the polarization to another random state, and I (or my poor students) have to go realign the system just because the lab temperature changed by five degrees.

Unpolarized light

Unpolarized light is the most common polarization state. *Unpolarized light* means that the electric field, whether rotating or linear, has no single defining state of oscillation among all the light waves present. A single light wave can't be unpolarized because that would mean that the electric field vector would change direction randomly, and as far as I know, that doesn't happen. Natural light, like the light from the sun or the light bulbs in your room, is unpolarized light. Figure 8-11 gives you an idea what unpolarized light looks like. Compare this figure with the other figures in this chapter, and you can see a clear difference between polarized and unpolarized light.

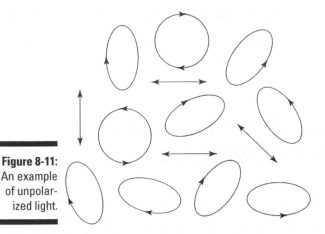

Figure 8-11:
An example
of unpolar-
ized light.

Unpolarized light is an important design characteristic for several applications that involve communication. Fiber-optic communication links often use unpolarized light to minimize the effect of temperature on error rates (see the preceding section for more on temperature's influence on experiments). Unpolarized light can play a big role in how effective wireless communication is, and you can use it to increase or decrease the security of the data transmitted.

Producing Polarized Light

Many different processes produce light (see Chapter 3), and they all produce unpolarized light when you're dealing with a large number of light waves. The following sections describe the basic processes that are able to convert unpolarized light, which is the state in which most light sources produce light, to polarized light.

Selective absorption: No passing unless you get in line

Selective absorption or *dichroism*, which occurs in materials where one linear polarized state is absorbed much more than the orthogonal linear polarized state, is the simplest way to produce linearly polarized light. Materials (typically polymers) that are made with oblong molecules are able to selectively absorb light based on the orientation of the electric field in the incident light. If the material has most of the molecules oriented with the long axes parallel to each other, it absorbs light whose electric field is parallel to the long axis because the electrons in the molecules move more freely along the long axis than the short dimension. However, the material doesn't absorb light whose electric field is parallel to the short dimension. The net effect is that one linear polarization state of the incident unpolarized light passes through such material while the other half of the light (with the orthogonal linear polarization state) is absorbed, resulting in the transmitted light being linearly polarized, as Figure 8-12 shows.

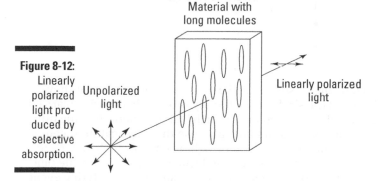

Figure 8-12: Linearly polarized light produced by selective absorption.

The process of selective absorption is very much dependent on the wavelength of the light. Materials that can absorb blue light often poorly absorb yellow or red light, and materials that absorb red light poorly absorb blue light. Converting all the wavelengths of sunlight to polarized light is very difficult, but lasers are another story. Generally, you can easily find a material that does a good job of polarizing the color of light emitted by the laser you have.

Scattering off small particles

Light is a transverse wave, so when it encounters small particles, atoms, or molecules that are smaller than the wavelength of the light, the light is *scattered off*, which is what happens when light bounces off atoms instead of reflecting (which happens with a surface — see Chapter 4 for more details). Scattering at 90 degrees causes the light to be linearly polarized because you can't have a component of the electric field in the direction of propagation before or after scattering. So when light scatters off small particles, the scattered light becomes linearly polarized, as you can see in Figure 8-13. (Check out the earlier section "Linear polarization" for more on this type of polarization.)

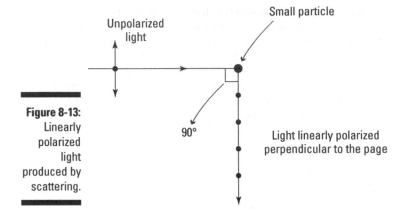

One reason polarized sunglasses can make objects stand out is that they reduce the irradiance of the sky because the sky is partially linearly polarized. Blue light is more readily scattered in the earth's atmosphere than the other wavelengths are, so the sky appears blue because the blue scattered light is directed to your eye from the sky. The blue light is only partially polarized because the atmosphere contains a lot of molecules, which scatter the blue light many, many times before it gets to your eye, and each scattering event often (but not always) changes the orientation of the polarization. Polarized glasses, as long as you orient them properly, can block some of the light from the sky more than they block light that reflects off of clouds and trees, giving you the impression that the objects stand out more.

Reflection: Aligning parallel to the surface

Light incident on a flat surface can reflect such that the reflected light is linearly polarized. For the reflected ray to be linearly polarized, you need a particular angle of incidence, called the *polarizing angle* or *Brewster's angle*. Figure 8-14 shows the arrangement by which this process happens.

Brewster's angle depends on the indexes of refraction of the materials light is traveling through or trying to enter. You can calculate this angle by using the following equation:

$$\theta_B = \tan^{-1}\left(\frac{n_2}{n_1}\right)$$

In this equation,

✔ θ_B is Brewster's angle or the angle of incidence that results in the reflected light being linearly polarized.

✔ n_1 is the index of refraction of the material the light starts out in.

✔ n_2 is the index of refraction of the material the light is trying to go into.

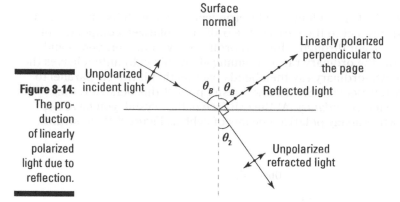

Figure 8-14: The production of linearly polarized light due to reflection.

The orientation of the polarized reflected light is linear and parallel to the surface. Because of dispersion (see Chapter 4), the materials' indexes of refraction depend on the wavelength of the light.

Polarized sunglasses are designed to block reflected light. Most often, they're designed to block the glare, which is the image of the sun reflecting off a surface like water, snow, ice, or the window or hood of an oncoming car. Because most of these surfaces are horizontal, polarized sunglasses are designed to block horizontally polarized light and allow vertically polarized light to pass through (see the earlier section "Linear polarization" for details on horizontally and vertically polarized light).

Birefringence: Splitting in two

Certain materials have two indexes of refraction, a characteristic known as *birefringence*. Each index of refraction corresponds to the same direction of propagation, but the light's linear polarization determines the specific index that the light waves experience in the material. The two linear polarization states are orthogonal to each other. Inside the material, light with electric field components parallel to one direction has one index, and the orthogonal component experiences a different value for the index of refraction. One orientation causes light to refract at an angle predicted by Snell's law. This index is the *ordinary index of refraction* and has the same value for all directions inside the material for that orientation of light. The other orientation can cause the light to refract at an angle not predicted by Snell's law. This index of refraction, the *extraordinary index of refraction,* doesn't have a fixed value because this index depends on the direction of travel in the material.

Light can follow one direction in birefringent materials where the ordinary and extraordinary indexes are the same. This direction is called the *optic axis* of the material. Light traveling parallel to this direction has the same index of refraction for all linearly polarized states.

If unpolarized light is incident on a birefringent crystal with its optic axis at an angle with respect to the surface, the ordinary polarized component of light passes straight through. The extraordinary ray, however, bends and separates from the ordinary ray, continuing along this path until it leaves the crystal. The extraordinary ray then bends back so that it travels parallel to the ordinary ray. Because Snell's law doesn't predict this behavior, the scientists call it *extraordinary*. At the exit face of the crystal, you have two separated, orthogonally polarized beams of light, as Figure 8-15 illustrates.

Figure 8-15:
An arrangement showing double refraction of unpolarized incident light.

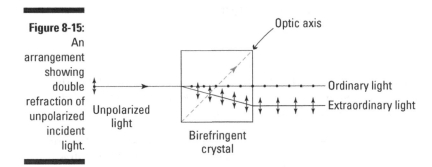

If you look at some words on a page through such a crystal (such as a piece of quartz from a science material supply store), you see two separated images that are orthogonally polarized.

This process of having two separated beams from one incident beam is usually referred to as *double refraction*. Although this phenomenon makes for a cool lecture demonstration, it's often a problem when designing optical systems that rely on using materials with birefringence.

As long as you know what to look for, you can avoid having double refraction appearing where you don't want it. Two separate light paths happen only if the light waves travel at an angle to the optic axis that isn't 0 degrees or 90 degrees. If the light travels parallel to the optic axis (0 degrees), only one index of refraction exists for all polarization states, so you don't have any beam separation. If the light waves travel at 90 degrees to the optic axis, the material has two indexes of refraction, but the light experiences no separation. This orientation is what you use to produce a phase delay between the two orthogonal polarization components. This phenomenon is one method used to change the polarization state of light (see Chapter 10).

Chapter 9

Changing Optical Polarization

. . . .

In This Chapter

▶ Introducing devices that can change the polarization of light

▶ Using equations to model the effect of polarization-changing devices

. . . .

*O*ptical polarization plays an important role in many optical systems that you may use. After light is polarized in some state (see Chapter 8 for details about how to do this), you may need to change that state for a variety of applications — for instance, like putting digital information on a laser beam. Most of the optical encoding procedures use the basic principles presented in this chapter to produce the data that is sent down fiber-optic cables. Polarization-changing devices also play a key role in 3-D movies. (Head to Part V for more on these applications.)

In this chapter, I present some basic optical devices that can change the polarization state of light. I also show a convenient notation to represent the various polarization states and the equations to determine the output of polarization-changing devices.

Discovering Devices that Can Change Optical Polarization

Some applications require a change in the polarization state, not just converting from unpolarized light to polarized light (I discuss the techniques that you can use to create polarized light in Chapter 8). In this section, I present some common devices that change the polarization state in a controlled, predictable fashion.

Dichroic filters: Changing the axis with linear polarizers

In Chapter 8, I present the idea behind *dichroism,* or selective absorption, which is based on the orientation of the electric field in the light wave relative to the structure in the material. Light with a linear component in one direction passes through the material, and the other orthogonal component is absorbed, making the transmitted light linearly polarized. Dichroism is one way to produce linearly polarized light out of incident unpolarized, or natural, light.

Because they produce linearly polarized light, dichroic filters are usually called *linear polarizers.* You can use linear polarizers to change the orientation of linearly polarized light, but at the expense of irradiance.

The transmission axis: Noting which orientation will pass through the filter

Linear polarizers can't change the polarization state; they can only make linearly polarized light. However, these polarizers can change the orientation of the linearly polarized light.

With dichroic filters, you need to know the orientation of linearly polarized light that the device will pass. The electric field orientation that passes through the dichroic filter is parallel to the *transmission axis* of the filter. Figure 9-1 shows the effect of a dichroic filter on initially unpolarized light.

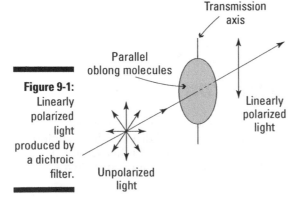

Transmission axis

Parallel oblong molecules

Linearly polarized light

Unpolarized light

Figure 9-1: Linearly polarized light produced by a dichroic filter.

When you buy a dichroic filter, the orientation of the transmission axis is usually noted with white or black marks that are opposite each other at the filter's edge or on its holder. If you send unpolarized light into a dichroic filter, the light that passes through will be linearly polarized oriented parallel to the transmission axis of the filter, whatever its orientation is, even if you rotate the filter.

If the initial light is unpolarized and you're monitoring the amount of light passing through, the irradiance you measure is half the incident level. For all possible orientations of the transmission axis, exactly half of the light passes through, regardless of the filter's orientation, because each orientation has an equal amount of light. If the initial light is linearly polarized, the irradiance of the transmitted light depends on the angle between the linearly polarized light and the transmission axis of the dichroic filter.

Finding the change in irradiance with Malus' law

You can use a series of linear polarizers to change the orientation of the linearly polarized light, but doing so costs you some energy. In many applications, you need to know the amount of irradiance each polarizer loses. *Malus' law* predicts how much irradiance will pass through each polarizer. This law is an idealization, because most dichroic filters don't work equally well for all wavelengths, but it gives you an idea about what happens.

Malus' law is used with linearly polarized light and linear polarizers and is

$$I = I_0 \cos^2 \theta$$

In this equation,

- ✔ I is the irradiance of the light that passes through the linear polarizer.

- ✔ I_0 is the irradiance of the linearly polarized light incident on the linear polarizer.

- ✔ θ is the angle between the orientation of the linearly polarized incident light and the transmission axis of the linear polarizer.

Look at the arrangement of two polarizers in Figure 9-2. Unpolarized light is incident on the first polarizer, whose transmission axis is vertical. The second polarizer is oriented such that its transmission axis is at 35 degrees. In a two-polarizer arrangement like this one, the second polarizer is often called an *analyzer*. Notice that in this arrangement, the dichroic filter has changed the orientation of the linearly polarized light to 35 degrees from the vertical. You may ask why I didn't just rotate the first polarizer to 35 degrees. For one thing, Malus' law requires at least two polarizers with incident unpolarized light. Also, you sometimes can't change the initial orientation of the linearly polarized light. Light from a laser may come out with a particular orientation that may not be good for your experiment.

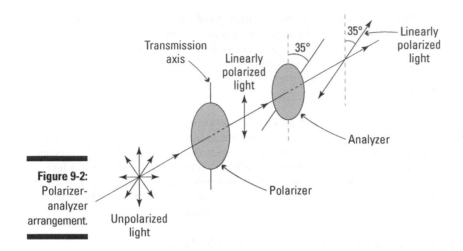

Figure 9-2:
Polarizer-
analyzer
arrangement.

If you want to calculate the final irradiance after the analyzer in Figure 9-2, you need to know the irradiance incident on the analyzer. To do that, you need to find the irradiance that passes through the first polarizer.

Malus' law doesn't work for the first polarizer because the incident light is unpolarized; θ is undefined for unpolarized light. So unless you know the wavelength of the light and the specific type of dichroic material you're using (so that you can look up the transmission characteristics from the manufacturer's specifications chart), use one-half the incident irradiance as the transmitted irradiance, which becomes the value you use for I_0 for the analyzer. (***Note:*** If the light is linearly polarized when it hits the first polarizer, you use the given value for the incident irradiance of light for I_0 and look at the angle between the incident linearly polarized light and the transmission axis of the polarizer for θ.)

For the situation presented in Figure 9-2, take the irradiance of the unpolarized light to be 4.2 watts per square meter. Because the initial light is unpolarized, you can't use Malus' law. Taking the transmitted irradiance as one-half the incident irradiance, the transmitted irradiance equals 2.1 watts per square meter after the first polarizer.

For the analyzer, you can use Malus' law because the light incident on the second polarizer is linearly polarized. The light through the first polarizer is oriented vertically because the transmission axis of the first polarizer is oriented that way. The analyzer is oriented with its transmission axis at 35 degrees from the vertical. θ is the angle between the orientation of the linearly polarized light and the transmission axis of the linear polarizer so θ is 35 degrees in this case. Using these values in Malus' law,

$$I = I_0 \cos^2 \theta$$
$$I = \left(2.1\text{W/m}^2\right)\cos^2\left(35°\right)$$

When you type this second equation in your calculator, you get 1.4 watts per square meter. Because the orientation of the linearly polarized light changed, the irradiance is lower.

To evaluate the $\cos^2$ function, make sure that you find the cosine of the angle first and then square that value. Don't square the angle before finding the cosine value.

Birefringent materials: Changing or rotating the polarization state

Unlike linear polarizers (see the preceding section), birefringent materials can rotate and change light's polarization state without significantly reducing the irradiance of the incident light. A *birefringent material* has two indexes of refraction. The direction that has the higher index of refraction is called the *slow axis*, and the other direction is called the *fast axis*.

Most of the birefringent materials that are commonly used in polarization-changing arrangements are crystals that have a special direction in them called the *optic axis*. Light that travels parallel to the optic axis experiences a single index of refraction (see Chapter 4 for more details), so it represents a unique direction in birefringent crystals. In fact, this direction is important to setting up polarization-changing devices correctly.

When light enters a birefringent crystal and travels perpendicular to the optic axis, you can describe it as the resultant of two orthogonal linear polarization states. One component has its electric field oscillating parallel to the optic axis, and the other component has its electric field oscillating perpendicular to the optic axis.

Figure 9-3 illustrates a typical arrangement for a birefringent crystal polarization-changing device.

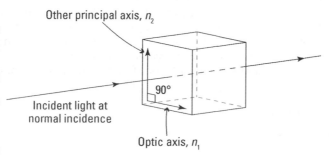

Figure 9-3:
Typical birefringent crystal indexes of refraction orientations.

Other principal axis, n_2

90°

Incident light at normal incidence

Optic axis, n_1

The light needs to travel perpendicular to the optic axis. In that case, only the electric-field component of the light is parallel to the optic axis. If light travels at an angle other than 90 degrees, two beams will emerge because of double refraction (see Chapter 8 for more information). Sometimes, this situation doesn't cause a problem, but sometimes it's the reason a birefringent application doesn't work.

Noting the phase change between components

If you want to use a birefringent crystal as a polarization-changing device, you need the orientation of linearly polarized light to *not* line up perfectly with either axis. For a birefringent crystal to function as a polarization-changer, you must have two components for the incident light so that each component travels through the crystal at different rates. This setup produces a phase change between the two components that then can change the polarization state or orientation, depending on how much of a phase change is introduced.

Figure 9-4 shows linearly polarized light that is oriented at 45 degrees from the slow axis. To see how the crystal will treat this light, you split the light into components parallel to each axis; one component parallel to the slow axis, the other one parallel to the fast axis, as shown in the figure.

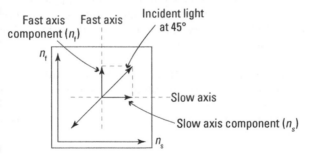

Figure 9-4: Linearly polarized light, with components, oriented at 45 degrees in a crystal.

For linearly polarized light, you can use a sine function to describe the behavior of the electric field as a function of time. Light that is linearly polarized has each component *in phase* with each other; the two components have no phase difference between them. This characteristic means that when the *x*-component of the electric field in Figure 9-4 reaches its maximum value, the *y*-component reaches its maximum value as well. The equations that describe each component's magnitude as a function of time at a point along the *z* axis — say, $z = 0$ — are given by the following:

$$E_x = E_{0x} \sin(\omega t)$$
$$E_y = E_{0y} \sin(\omega t)$$

In these equations,

- ✔ E_x is the instantaneous electric field in the x-direction.
- ✔ E_{0x} is the electric field amplitude in the x-direction.
- ✔ ω is the angular frequency of the electric field defined as 2π times the frequency, *f.*
- ✔ *t* is time.
- ✔ E_y is the instantaneous electric field in the y-direction.
- ✔ E_{0y} is the electric field amplitude in the y-direction.

The relationship between the components keeps the resultant polarization state linear.

If the light travels through a birefringent crystal, perpendicular to the optic axis with linear polarization at some angle relative to the fast axis other than 0 or 90 degrees, each component travels with a different speed through the material. Therefore, you get a *phase delay* between the components when they get to the end of the crystal. That is, the waves don't line up crest with crest but rather have a shift between their crests. As long as you know the values for the two indexes of refraction in the material, you can use the following formula to calculate what the phase delay will be:

$$\Gamma = \frac{2\pi}{\lambda}d|n_s - n_f|$$

In this equation,

- ✔ Γ is the phase delay (in radians!) between the two orthogonal linearly polarized components in the crystal.
- ✔ λ is the vacuum wavelength of the light incident on the crystal.
- ✔ *d* is the actual thickness of the crystal that the light is traveling through.
- ✔ n_s is the larger index of refraction corresponding to the slow axis in the crystal.
- ✔ n_f is the smaller index of refraction corresponding to the fast axis in the crystal.

Depending on which orientation corresponds to the fast axis, the electric field component that oscillates parallel to the fast axis travels faster through the material than the component whose field is oscillating parallel to the slow axis does. Therefore, the fast component speeds up (or is *phase advanced*) relative to the slow component by the amount Γ. In the situation shown in

Figure 9-4, the slow axis is horizontal, so the equations for the electric field are the following:

$$E_x = E_{0x} d \sin(\omega t)$$
$$E_y = E_{0y} \sin(\omega t + \Gamma)$$

As the two components leave the crystal, they're no longer in phase with each other.

Except for certain thicknesses of crystal, the light is no longer linearly polarized. If the thickness causes Γ to be an integral number times π, the light remains linearly polarized. Otherwise, the light is generally elliptically polarized. (Check out Chapter 8 for more on the causes of elliptical polarization.)

Changing the phase with wave plates

You can use birefringent materials to either rotate or change the polarization state. After you choose your materials, the only other design variable is the thickness of the material that the light travels through.

Two very common optical devices use birefringent materials to rotate or change the polarization state: the quarter-wave plate and the half-wave plate:

- **Quarter-wave plate:** A *quarter-wave plate* is a polarization-changing device made from a birefringent crystal cut with a thickness that produces a phase delay of $\pi/2$ between the two components in the material. It's the only commonly made polarization-changing device that changes the type of polarization and not the orientation of the polarization state. If the light is linearly polarized and oriented at 45 degrees to either axis, the resulting light is circularly polarized light. If the light is at any other angle (besides 0 or 90 degrees), the resulting light is elliptically polarized. *Note:* Because of dispersion and the fact that the wavelength appears in the equation for Γ, a quarter-wave plate is designed for a specific wavelength and doesn't work as a quarter-wave plate if you use light with a different wavelength. See Chapter 4 for more on dispersion.

- **Half-wave plate:** The half-wave plate is a polarization-state rotator. The thickness of the birefringent material is set such that the phase difference between the two components is π. This phase difference doesn't change the polarization state, but it does rotate the polarization state. If light is polarized either linearly or elliptically (you can't tell the difference if a circle is rotated), the orientation measured relative to one of the axes doubles by that angle. If linearly polarized light is incident on a half-wave plate at an angle of 25 degrees relative to the horizontal, it emerges at an angle of –25 degrees (25 degrees below the horizontal [x-axis]). If elliptically polarized light is incident with its major axis oriented at 40 degrees relative to the horizontal, it emerges with its major axis oriented at –40 degrees. Half-wave plates fall victim to the same wavelength dependence as quarter-wave plates.

Rotating light with optically active materials

A special class of materials functions like the half-wave plate (polarization rotator) without the orientational dependence of the input light. These materials have a helical structure that causes the electric field of the incident light to be rotated, but not because of a phase delay between orthogonal components like the birefringent materials earlier in this chapter. Rather, these materials are called *optically active materials* because of their ability to rotate the polarization state of the incident light. The amount of rotation depends on the thickness of the materials.

Jones Vectors: Calculating the Change in Polarization

Many different configurations of materials produce a wide variety of changes in polarization state. In this section, I show you a convenient way to describe polarization states mathematically and how to calculate the change that the materials cause to a given polarization state. This process involves matrices, but it's rather straightforward. (Chapter 2 provides a refresher on working with matrices.)

Because light is a transverse wave, the orientation of the electric field is specified in two dimensions, usually x and y. This setup means that the matrices involved have only two or four elements in them, as I show in the following sections.

Representing the polarization state with Jones vectors

Jones vectors are simple 2 x 1 matrices that allow you to easily describe any polarization state. Jones vectors serve as the input into devices that will change the polarization and serve as the way to see what the output of a particular device or series of devices will be. Table 9-1 gives the Jones vector representations for the different polarization states. Note that unpolarized light has no Jones vector representation.

Table 9-1	Polarization States and Jones Vectors	
Polarization State	*Orientation*	*Jones Vector*
Linear Polarization: $\Gamma = m\pi$	General orientation measured at an angle α with respect to the x-axis.	$\begin{bmatrix} \cos\alpha \\ \sin\alpha \end{bmatrix}$
	Vertical ($\alpha = 90°$.)	$\begin{bmatrix} 0 \\ 1 \end{bmatrix}$
	Horizontal ($\alpha = 0°$.)	$\begin{bmatrix} 1 \\ 0 \end{bmatrix}$
	$\alpha = +45°$.	$\frac{1}{\sqrt{2}}\begin{bmatrix} 1 \\ 1 \end{bmatrix}$
	$\alpha = -45°$.	$\frac{1}{\sqrt{2}}\begin{bmatrix} 1 \\ -1 \end{bmatrix}$
Circular Polarization: $\Gamma = (2m + \frac{1}{2})\pi$	Left (when viewed with the light traveling toward you).	$\frac{1}{\sqrt{2}}\begin{bmatrix} 1 \\ i \end{bmatrix}$
	Right (when viewed with the light traveling toward you).	$\frac{1}{\sqrt{2}}\begin{bmatrix} 1 \\ -i \end{bmatrix}$
Elliptical Polarization: $\Gamma = (m + \frac{1}{2})\pi$	Left (when viewed with the light traveling toward you). Major axis is parallel to the axis with the largest amplitude, E_{0x} or E_{0y}.	$\frac{1}{\sqrt{E_{0x}^2 + E_{0y}^2}}\begin{bmatrix} E_{0x} \\ iE_{0y} \end{bmatrix}$
	Right (when viewed with the light traveling toward you). Major axis is parallel to the axis with the largest amplitude, E_{0x} or E_{0y}.	$\frac{1}{\sqrt{E_{0x}^2 + E_{0y}^2}}\begin{bmatrix} E_{0x} \\ -iE_{0y} \end{bmatrix}$

Polarization State	Orientation	Jones Vector
Elliptical Polarization: $\Gamma \neq (m + \frac{1}{2})\pi$ or $m\pi$	Left (when viewed with the light traveling toward you). Major axis is oriented at an angle α with respect to the x-axis.	$\dfrac{1}{\sqrt{A^2+B^2+C^2}}\begin{bmatrix} A \\ B+iC \end{bmatrix}$
		$\alpha=\dfrac{1}{2}\tan^{-1}\left(\dfrac{2A\sqrt{B^2+C^2}\cos\beta}{A^2-B^2-C^2}\right)$
		$\beta=\tan^{-1}\left(\dfrac{C}{B}\right)$
	Right (when viewed with the light traveling toward you). Major axis is oriented at an angle α with respect to the x-axis.	$\dfrac{1}{\sqrt{A^2+B^2+C^2}}\begin{bmatrix} A \\ B-iC \end{bmatrix}$
		$\alpha=\dfrac{1}{2}\tan^{-1}\left(\dfrac{2A\sqrt{B^2+C^2}\cos\beta}{A^2-B^2-C^2}\right)$
		$\beta=\tan^{-1}\left(\dfrac{C}{B}\right)$

In Table 9-1 for the elliptical polarization state, A is equal to the amplitude E_{0x}, B is the real part of the amplitude E_{0y}, and C is the imaginary part of the amplitude E_{0y}, such that E^2_{0y} has a magnitude of $B^2 + C^2$. The listing of these variables is so that you can determine the orientation of the major axis (the angle α) given the Jones vector for elliptically polarized light.

The information in Table 9-1 provides a list of all the possible polarization states. The general elliptical polarization states are the nastiest looking expressions, but they're probably the most common. Most optical engineers design optical systems that don't produce these kinds of states, but you may sometimes get such a polarization state. When that happens during preliminary calculations, you can correct your design before you have to deal with this problem in an actual device.

Jones matrices: Showing how devices will change polarization

With a convenient matrix representation of polarization states, you can model polarization-changing devices so that you can predict the output of a particular arrangement of polarization changing devices.

You need to worry about three basic, commonly used polarization-changing devices only:

- **Linear polarizer:** The linear polarizer (which I discuss in the earlier section "Dichroic filters: Changing the axis with linear polarizers") is the only device of the three that can do anything measureable with incident unpolarized light. You can use the polarizer to rotate the linear polarization state, as long as you don't try to do so with an angle of 90 degrees.

- **Phase retarder:** *Phase retarders* are birefringent materials that can change the phase delay between two orthogonal, linearly polarized components. This device is the first that I've mentioned that can actually change the polarization state. The quarter- and half-wave plates I cover earlier in the chapter are two very common phase retarders in optics. Flip to Chapter 8's elliptical polarization discussion to find out more about the effect of the phase delay.

- **Rotator:** This device can't change the polarization state; it rotates the polarization state, whether linear or elliptical, only. The half-wave plate is a phase retarder that functions like a rotator. Optically active materials can also be used as rotators, but without worrying about the orientation of the material relative to incident polarization state.

To formulate a procedure that helps predict the final polarization state of the light after it travels through a single device or any combination of them, you need a mathematical operation that can model the effect of the three types of devices. With Jones vectors representing the polarization state of the incident light (see the preceding section), you can use matrices to represent the effect of the element on the input polarization state.

Table 9-2 presents the matrices that perform the listed operation on the input polarization states. These matrices are 2 x 2 matrices because they represent operations or transformations to be performed on the input polarization state, represented by the Jones vector.

Table 9-2 Polarization-Changing Devices and Jones Matrices

Polarization Changing Device	Device Orientation	Jones Matrix
Linear Polarizer	Transmission axis oriented at an angle θ with respect to the x-axis.	$\begin{bmatrix} \cos^2\theta & \sin\theta\cos\theta \\ \sin\theta\cos\theta & \sin^2\theta \end{bmatrix}$
	Transmission axis parallel to the x-axis.	$\begin{bmatrix} 1 & 0 \\ 0 & 0 \end{bmatrix}$
	Transmission axis parallel to the y-axis.	$\begin{bmatrix} 0 & 0 \\ 0 & 1 \end{bmatrix}$
	Transmission axis oriented at 45 degrees with respect to the x-axis.	$\dfrac{1}{2}\begin{bmatrix} 1 & 1 \\ 1 & 1 \end{bmatrix}$
Phase Retarder (Wave plate)	General phase delay.	$\begin{bmatrix} e^{i\phi_x} & 0 \\ 0 & e^{i\phi_y} \end{bmatrix}$
	Quarter-wave plate ($\Gamma=\pi/2$), slow axis parallel to the y-axis.	$e^{-i\pi/4}\begin{bmatrix} 1 & 0 \\ 0 & i \end{bmatrix}$
	Quarter-wave plate ($\Gamma=\pi/2$), slow axis parallel to the x-axis.	$e^{i\pi/4}\begin{bmatrix} 1 & 0 \\ 0 & -i \end{bmatrix}$
	Half-wave plate ($\Gamma=\pi$) slow axis parallel to the y-axis.	$e^{-i\pi/2}\begin{bmatrix} 1 & 0 \\ 0 & -1 \end{bmatrix}$
	Half-wave plate ($\Gamma=\pi$) slow axis parallel to the x-axis.	$e^{i\pi/2}\begin{bmatrix} 1 & 0 \\ 0 & -1 \end{bmatrix}$
Rotator	Rotate the incident polarization state by an angle γ.	$\begin{bmatrix} \cos\gamma & -\sin\gamma \\ \sin\gamma & \cos\gamma \end{bmatrix}$

Matrix multiplication: Predicting how devices will affect incident light

With both Jones vectors and Jones matrices, you have all of the pieces you need to predict how a polarization-changing system will affect incident light. To find the output of a polarization-changing system, just follow these steps:

1. **Determine the initial polarization state of the light.**

   If the light is polarized, you're ready for Step 2. If the incident light is unpolarized, check to see whether the first element is a linear polarizer; if so, you know you have linear polarization oriented parallel to the transmission axis of the polarizer. If the device isn't a linear polarizer, doing the calculation doesn't help you because the effect will be undetectable.

2. **Generate the appropriate Jones vector for the polarization state that you have, using the information in Table 9-1.**

3. **Identify the polarization-changing devices and find their Jones matrices from Table 9-2.**

   If you identify only one matrix, you can go directly to Step 5.

4. **Create the system matrix.**

   If Step 3 yields more than one matrix, place each matrix right to left (this order is very important!), with the first matrix on the far right being for the device that the light encounters first. Calculate the new matrix by multiplying all the matrices together. This new matrix is the *system matrix.*

5. **Place the Jones vector to the right of the Jones matrix (or the system matrix) and multiply the two.**

6. **Simplify the resulting Jones vector by dividing out common factors from each component and identify the polarization state, using the information in Table 9-1.**

   The Jones vector you generate with the last device is the polarization state of the light at the output of the system.

When you look at some of the matrices in Tables 9-1 and 9-2, you may notice some numbers out in front of the matrices. These numbers are necessary only if you need to calculate the irradiance of the light that passes through; they don't affect the polarization state. For the most part, irradiance calculations involving Jones vectors and matrices are done for academic exercises; these calculations don't account for any of the absorption and reflection losses that occur in the real world.

Doing calculations with a single polarization-changing device

Suppose you have a system that begins with a linear polarizer with its transmission axis at 37 degrees to the horizontal (x-axis) on the input, followed by a quarter-wave plate with its slow axis horizontal (parallel to the x-axis). Assume that the incident light is unpolarized, and you want to find the final polarization state of light that exits.

You have unpolarized light, so in order for the calculations to mean anything, you need to see whether the next device in the system is a linear polarizer. In this situation, it is, so the light that passes through that first polarizer is linearly polarized light oriented parallel to the transmission axis of the polarizer. The transmission axis is oriented at 37 degrees to the horizontal, so you look at Table 9-1 to find the Jones vector for that type of light (subbing in the orientation angle), which is

$$\begin{bmatrix} \cos 37° \\ \sin 37° \end{bmatrix} = \begin{bmatrix} 0.7986 \\ 0.6018 \end{bmatrix}$$

Now you identify the Jones matrix for the next polarization-changing device, which is the quarter-wave plate in this situation. Looking at Table 9-2, you find the matrix for a quarter-wave plate with its slow axis parallel to the x-axis is

$$e^{i\pi/4} \begin{bmatrix} 1 & 0 \\ 0 & -i \end{bmatrix}$$

Because you're interested in determining only the polarization state (linear, circular, or elliptical) and not the irradiance, you don't need the phase term in front of the matrix, so you can get rid of it as follows and then multiply the Jones vector by this Jones matrix.

$$\begin{bmatrix} 1 & 0 \\ 0 & -i \end{bmatrix} \begin{bmatrix} 0.7986 \\ 0.6018 \end{bmatrix} = \begin{bmatrix} 0.7986 \\ -i0.6018 \end{bmatrix}$$

Find the matrix in Table 9-1 that matches the matrix you just generated. Because the y-component contains an i, the polarization is a rotating one and is therefore circular or elliptical; the two components have different magnitudes, so the polarization state must be elliptical. Because the x-component is larger than the y-component, the major axis is oriented parallel to the x-axis (horizontal). The $-i$ in the y-component means that the light is right elliptically polarized; because the quarter-wave plate is the last element in your system, this state is also the output polarization state.

To save yourself some unnecessary work in these kinds of problems, make sure that any linear incident polarization state isn't parallel to either the fast or slow axis of any phase retarder. If you put horizontal or vertical linearly polarized light with any phase retarder with a horizontal axis, nothing happens. The input and output polarization states are exactly the same. In this situation, you don't need to waste your time calculating anything (unless your teacher tells you to); the birefringent material won't do anything to the polarization state because the material has only one component. This situation doesn't apply to a rotator that uses an optically active material because the rotator isn't a phase retarder. Optically active rotators work for any orientation of polarized light.

Using multiple polarization-changing devices

Sometimes you may have a problem that gives you two or more polarization-changing devices in a row. Using matrices allows you to use a shortcut: Because all the effects of the devices are represented by a 2-x-2 matrix, you can account for the effect of several devices by multiplying the matrices together, getting a new system Jones matrix that includes all the individual devices' effects. However, you must make certain that you multiply the matrices together in the right order as I note earlier in the section for this technique to work correctly.

If the last device in a polarization-changing system is a linear polarizer, you don't need to do any calculations. The point of the calculations in this chapter is to determine the polarization state of the output light. When the last element is a linear polarizer, the polarization is linear and oriented parallel to the transmission axis of the polarizer, regardless of the number of polarization-changing devices that come before it. So check this possibility first before you start crunching matrices together.

Suppose you have a system that begins with a linear polarizer with its transmission axis oriented at 22.5 degrees from the horizontal, a half-wave plate with its slow axis horizontal, and a quarter-wave plate with its slow axis vertical. Assume as well that you're working with unpolarized incident light. Because you have two polarization-changing elements, the half-wave plate and the quarter-wave plate, you need to compute a new system matrix. The incident light is unpolarized, so you don't need to include the linear polarizer in the system matrix because it creates the linearly polarized light that you need to get the system to work. Using the information in Table 9-2, you get the following system matrix:

$$\begin{bmatrix} 1 & 0 \\ 0 & i \end{bmatrix} \begin{bmatrix} 1 & 0 \\ 0 & -1 \end{bmatrix}$$

Notice that I place the matrices in the reverse order that the light goes through the elements. Now multiply the matrices together, starting with the two matrices on the far right. In this situation, you have only two matrices, so you can multiply them to get the system Jones matrix:

$$\begin{bmatrix} 1 & 0 \\ 0 & -i \end{bmatrix}$$

Put this matrix with the initial polarized state, which is linear oriented at 22.5 degrees above the horizontal to get

$$\begin{bmatrix} \cos 22.5° \\ \sin 22.5° \end{bmatrix} = \begin{bmatrix} 0.9239 \\ 0.3827 \end{bmatrix}$$

You put the Jones vector with the system Jones matrix and find

$$\begin{bmatrix} 1 & 0 \\ 0 & -i \end{bmatrix}\begin{bmatrix} 0.9239 \\ 0.3827 \end{bmatrix} = \begin{bmatrix} 0.9239 \\ -i0.3827 \end{bmatrix}$$

Comparing vector to the Jones vectors in Table 9-1, you find that the output polarization is right elliptically polarized with its major axis parallel to the horizontal.

Chapter 10

Calculating Reflected and Transmitted Light with Fresnel Equations

• •

In This Chapter

▶ Examining important quantities in reflectance and transmittance of light

▶ Reflecting on external and internal reflection

• •

*T*he laws of reflection and refraction (see Chapter 3) are useful for finding where an image may appear, but they don't tell you how much light is reflected or transmitted. In many applications, you need to determine the amounts of light present in order to make the device function properly (for example, to make an image that's bright enough for you to see or to accurately send data to, say, your HDTV).

In this chapter, I define the optical terms for the amount of reflected and transmitted light as well as show you what factors play a role in determining the amount of light in each case. I also let you in on some interesting effects that happen with different types of reflection. *Remember:* In designing optical systems, you must consider the situations highlighted in this chapter to make the devices function properly.

Determining the Amount of Reflected and Transmitted Light

When light encounters a boundary between two different materials (a marker characterized by a change — however small — in the index of refraction) some of the light is transmitted and some is reflected. The laws of reflection and refraction in Chapter 4 provide you with equations so that you can find the directions that the light will follow, but they tell you nothing about how much of the incident light follows each path. In this section, I show you the quantities you use to determine the amount of light in each path. First, though, I define some terms used to talk about the amounts of light.

Transverse modes: Describing the orientation of the fields

The first thing you need to define is the orientation of the electric and magnetic fields. Although the orientation of the electric field is usually described by polarization (see Chapter 8), you need to look at the fields relative to how they're striking the surface when you want to find how much light follows each path. The equations you use to determine the amplitudes of the fields in the reflected and refracted waves are determined from boundary conditions for electric and magnetic fields.

The incident, reflected, and refracted rays all lie in a plane called the *plane of incidence*. In optics, the orientations of the fields are described as modes relative to the plane of incidence. When determining the amount of light carried by the reflected and refracted waves, you need to know the orientation of the fields because the amount of light is different for each orientation.

You can separate all the possible orientations of fields into two perpendicular components, like you can all vector quantities; you just need to find the directions that are most helpful to you in your particular situation. When looking at the amount of light carried by the various waves, you look at the field that's perpendicular to the plane of incidence.

✔ If the incident light has its electric field oriented perpendicular to the plane of incidence, the mode of the light is referred to as the *transverse electric* (TE) *mode*.

✔ If the incident light has its magnetic field oriented perpendicular to the plane, the mode of the light is referred to as the *transverse magnetic* (TM) *mode*.

Figure 10-1 shows what these two modes look like, with the plane of incidence being parallel to the page. The fields that are perpendicular to the page are indicated by dots (representing arrows oscillating perpendicular to the page). The TE mode is shown in the left diagram in Figure 10-1, and the TM mode is shown on the right.

The TE and TM modes are the two extreme cases; any arbitrary orientation of fields in the incident light can be made out of these two modes, just like a two-dimensional vector can be made out of two orthogonal components. When designing optical systems, an optical engineer usually chooses one of these orientations to work with to keep the problem as simple as possible.

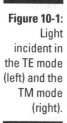

Figure 10-1:
Light
incident in
the TE mode
(left) and the
TM mode
(right).

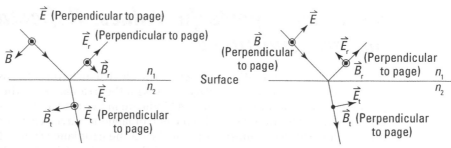

Defining the reflection and transmission coefficients

One fundamental quantity used to determine the amount of light in the reflected beam is the reflection coefficient. In Chapter 3, I talk about light being an electromagnetic wave and the wave carrying energy in its electric and magnetic fields. If the amount of energy in the wave decreases, the amplitudes of the field quantities reduce. The *reflection coefficient* gives the fraction of the incident electric field amplitude that appears in the reflected beam, mathematically defined as

$$r = \frac{E_r}{E_0}$$

In this equation,

- ✔ r is the reflection coefficient.
- ✔ E_0 is the electric field amplitude of the incident light.
- ✔ E_r is the electric field amplitude of the reflected light.

A similar quantity for the amount of light in the transmitted (refracted) beam is called the *transmission coefficient.* The transmission coefficient gives the fraction of the incident electric field amplitude that appears in the transmitted beam, defined mathematically as

$$t = \frac{E_t}{E_0}$$

In this equation,

- ✔ t is the transmission coefficient.
- ✔ E_0 is the electric field amplitude of the incident light.
- ✔ E_t is the electric field amplitude of the transmitted light.

Using more powerful values: Reflectance and transmittance

Because no instrumentation can detect the electric field directly, as a matter of practice, a more useful quantity deals with the irradiance, or the power per unit area carried by the light. As I indicate in Chapter 3, the Poynting vector represents the rate at which electromagnetic waves carry energy through an area, so quantities that deal with the irradiance (something you can measure) are much more useful. The irradiance is proportional to the square of the electric field (see Chapter 3 for the details). In terms of reflected and transmitted light, the percentage of the incident irradiance that appears in the reflected beam is called the *reflectance* and represented by the following equation:

$$R = r^2 = \left(\frac{E_r}{E_0} \right)^2$$

The percentage of the incident irradiance that appears in the transmitted beam is the *transmittance*. The transmittance expression is more complicated because the index of refraction changes, which means that the direction and the apparent speed that the light travels in the material change, too:

$$T = \left(\frac{n_2}{n_1} \right) \left(\frac{\cos \theta_t}{\cos \theta_i} \right) t^2$$

You can use conservation of energy with both the reflection and transmission coefficients and the reflectance and transmittance. The equations for the transmitted quantities are typically more complicated, but the principle of conservation of energy can make them easy to calculate if you find the appropriate reflection quantity first. The coefficient terms are split into two conditions based on the mode of the light, transverse electric or transverse magnetic (TE or TM; see the preceding section):

TE: $t = r + 1$

TM: $\left(\frac{n_2}{n_1} \right) t = r + 1$

For the reflectance and transmittance, you have one expression,

$$R + T = 1$$

All three of these expressions are just alternative ways to say that all the energy delivered to the surface must be carried away. The relationships I present in this chapter are idealized because the equations don't account for the scattering losses typically experienced at the surface in actual applications. However, these idealized relationships still do a very good job of predicting what will happen in a general sense.

The Fresnel equations: Finding how much incident light is reflected or transmitted

REMEMBER

The *Fresnel equations* are four equations used to determine how much light appears in the reflected and transmitted waves as a fraction of the incident light. They're derived from the boundary conditions for electric and magnetic fields as they encounter a surface. Here they are:

$$r_{TE} = \frac{\cos\theta - \sqrt{n^2 - \sin^2\theta}}{\cos\theta + \sqrt{n^2 - \sin^2\theta}}$$

$$r_{TM} = \frac{n^2\cos\theta - \sqrt{n^2 - \sin^2\theta}}{n^2\cos\theta + \sqrt{n^2 - \sin^2\theta}}$$

$$t_{TE} = \frac{2\cos\theta}{\cos\theta + \sqrt{n^2 - \sin^2\theta}}$$

$$t_{TM} = \frac{2n\cos\theta}{n^2\cos\theta + \sqrt{n^2 - \sin^2\theta}}$$

In these equations,

- r is the reflection coefficient.
- θ is the incident angle of the light.
- n is the relative index of refraction defined as n_2/n_1.
- t is the transmission coefficient.

These equations give the respective reflection and transmission coefficients, which you can use to determine the percentage of the incident power that appears in the reflected and transmitted beams.

Suppose that you want to find the amount of light reflected and transmitted (for both TE and TM modes) for light incident at 20 degrees on a piece of glass with an index of refraction of 1.50. In a practical sense, the reflectance and transmittance are the most useful quantities, so that's what I show you how to find.

1. **Calculate the coefficients of reflection for both modes.**

   Use the r_{TE} Fresnel equations:

   $$r_{TE} = \frac{\cos\theta - \sqrt{n^2 - \sin^2\theta}}{\cos\theta + \sqrt{n^2 - \sin^2\theta}}$$

   $$= \frac{\cos 20° - \sqrt{(1.5)^2 - \sin^2 20°}}{\cos 20° + \sqrt{(1.5)^2 - \sin^2 20°}} = -0.2170$$

$$r_{TM} = \frac{n^2 \cos\theta - \sqrt{n^2 - \sin^2\theta}}{n^2 \cos\theta + \sqrt{n^2 - \sin^2\theta}}$$

$$= \frac{(1.5)^2 \cos 20° - \sqrt{(1.5)^2 - \sin^2 20°}}{(1.5)^2 \cos 20° + \sqrt{(1.5)^2 - \sin^2 20°}} = 0.1829$$

You may notice the negative sign in the r_{TE} number. I discuss its significance in "Surveying Special Situations Involving Reflection and the Fresnel Equations" later in the chapter.

2. **Find the transmittance for both modes.**

You can use the Fresnel equations to find the transmission coefficient, but using conservation of energy makes the calculations much easier than using the expressions for the transmission coefficients and transmittances.

Together, reflectance and transmittance have to represent 100 percent of the incident light: $R + T = 1$. First, you find the reflectance, using the reflectance equation from the earlier section "Using more powerful values: Reflectance and transmittance"

$$R_{TE} = r_{TE}^2 = 0.0471$$

$$R_{TM} = r_{TM}^2 = 0.03345$$

Reflectances and transmittances are typically expressed as percentages, so R_{TE} is 4.71 percent and R_{TM} is 3.35 percent. Using these values with the conservation of energy principle, you can easily find the transmittances through simple subtraction:

$$T_{TE} = 100\% - R_{TE} = 95.29\%$$

$$T_{TM} = 100\% - R_{TM} = 96.65\%$$

If the incident light is unpolarized, half the irradiance is in the TE mode and the other half is in the TM mode. If you measure the irradiance and don't care about the polarization, the total reflectance and transmittance is the average of the TE and TM modes. In this particular example, a total of 4.03 percent of the incident light is reflected, and 95.97 percent is transmitted.

Surveying Special Situations Involving Reflection and the Fresnel Equations

The reflection and transmission coefficients vary with the angle of incidence, so they show some interesting situations that you can use in variety of applications. Although the index of refraction (*relative index* of refraction in the Fresnel equations) depends on wavelength, this dependence changes only the particular values for the angles of incidence.

In this section, I show you some of the special situations that arise regardless of the wavelength of the light involved. The special situations, such as total internal reflection (see Chapter 4) and linearly polarized reflected light (see Chapter 8), occur with two different types of reflection: external and internal:

- *External reflections* happen when n_2 is greater than n_1 or the relative index of refraction, *n,* is greater than 1.
- *Internal reflections* occur when n_1 is greater than n_2 or the relative index of refraction, *n,* is less than 1.

You can observe a general characteristic of external reflections if you look at the values for r_{TE} as a function of the incident angle. The r_{TE} values are negative for all values of incident angles. This negative designation doesn't affect the amount of light present in the reflected beam, but it does mean that the light wave is shifted by half a wavelength (or, in terms of phase, a shift of π) after it bounces off the surface. This change plays a major role in thin film interference applications, which I discuss in Chapter 12.

Striking at Brewster's angle

If you make a graph of r_{TM} as a function of the angle of incidence, you find an angle (*Brewster's angle,* or the *polarizing angle*) where the coefficient is zero. In Chapter 3, I talk about this angle with Snell's law, but this different perspective shows from fundamental electricity and magnetism principles that the reflected light is linearly polarized.

Using the reflection coefficients, you can see that the reflected light is polarized in the TE mode because r_{TE} isn't zero at this angle for light incident at Brewster's angle.

Reflectance at normal incidence: Coming in at 0 degrees

A very common occurrence in optics is where the light hits a surface at *normal incidence,* which means that the angle of incidence is zero degrees (in other words, the light is traveling perpendicular to the surface). Many laboratory setups and optical devices use light at normal incidence to the optical elements.

If you look at the reflection coefficients for the TE and TM modes earlier in the chapter, you may notice that they become equivalent expressions when θ is 0 degrees. As a matter of practice, reflectance is a more useful expression and therefore is rather useful when estimating energy losses in an optical system. It's given by

$$R = \left(\frac{1-n}{1+n}\right)^2$$

Using this expression, you can see the common value of 4 percent reflection off a single surface of typical glass from air (n = 1.50) at normal incidence.

Reflectance at glancing incidence: Striking at 90 degrees

If you look at the Fresnel equations and let the incident angle equal 90 degrees, you notice that both reflection coefficients go to –1.0 and the transmission coefficients go to zero. This value means that all the light is reflected off the surface and none of it transmitted through the surface.

When the incident angle approaches 90 degrees, you have what's called *glancing incidence*. All materials reflect nearly all the incident light when the light is at glancing incidence. Even materials that don't reflect much light, like the cover of this book, show the image of the light bulb in the ceiling if you hold them so that you're nearly looking along the surface toward the light bulb.

Exploring internal reflection and total internal reflection

With internal reflection, the light starts in the material with the larger index of refraction. Internal reflection shows some of the same characteristics as external reflection. In particular, the TM component goes to zero at Brewster's angle; the TE component isn't zero, making the reflected light polarized just like in the external reflection case except for a different value of the angle of incidence.

A big difference occurs in both reflection coefficients as you increase the angle of incidence past Brewster's angle: Both reach a value of 1 long before an incident angle of 90 degrees. Both reflection coefficients reach one when the term under the radical becomes zero — when $\sin\theta$ equals n. Both coefficients have a value of 1 when θ is at the critical angle, resulting in the incident light being totally internally reflected.

As the angle of incidence increases beyond this critical angle, the term under the radical becomes negative because $\sin\theta$ is greater than n, so both reflection coefficients are *complex* (contain real and imaginary parts). When this increase happens, the magnitude is still 1, indicating that all light incident on a surface undergoing internal reflection is totally internally reflected — no light goes through to the other material. This major phenomenon is used with fiber optics, which I talk about in more detail in Chapter 16.

Frustrated total internal reflection: Dealing with the evanescent wave

In the preceding section, I mention that no light enters the second material as long as the incident light has an angle of incidence greater than the critical angle. However, if you look very close to the surface where the light hits, you find a small, exponentially decaying wave in the second material. If you place another material close to, but not touching, the first boundary, the second material can pick up and propagate the wave away. This wave that passes into the second material is called an *evanescent wave*.

If this evanescent wave is intercepted with a high-index material before its amplitude is zero, the wave continues on in a process called *frustrated total internal reflection*. (No, that's not what you do during a midlife crisis.) The energy from this wave comes from the incident wave, so when total internal reflection is frustrated, the energy in the reflected beam decreases by the amount carried away in the evanescent wave. If the process isn't frustrated, the energy returns to the reflected beam so that the reflected beam experiences no energy loss.

The fiber-optics industry uses the phenomenon of frustrated total internal reflection quite a bit to control the amount of light in different parts of the fiber network (head to Chapter 16 for more information about this application). This phenomenon is also used in devices called *beam splitters* that are designed to (surprise!) split laser beams in two different directions.

A common use of frustrated total internal reflection utilizes this effect between two right-angle prisms by carefully controlling the space between the two prisms to transmit a desired percentage of the incident light and reflect the rest. This arrangement is one form of a beam splitter, known as a *cube beam splitter;* you can see one in Figure 10-2.

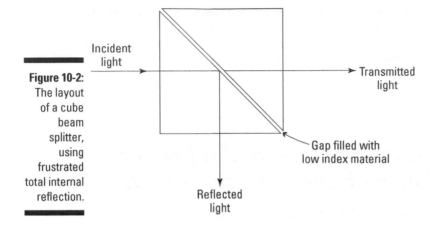

Figure 10-2:
The layout of a cube beam splitter, using frustrated total internal reflection.

Incident light

Transmitted light

Gap filled with low index material

Reflected light

The distance over which the evanescent wave exists is usually very small. For the evanescent wave, you can define a rate at which the energy dissipates as a function of distance, called the *absorption coefficient*, α. This coefficient is defined as

$$\alpha = \frac{2\pi n_2}{\lambda_0} \sqrt{\frac{\sin^2 \theta}{n^2} - 1}$$

In this equation,

- α is the absorption coefficient.
- λ_0 is the vacuum wavelength of the light.
- θ is the angle of incidence.
- n is the relative index of refraction, equal to n_2/n_1.

Evanescent waves have a standard parameter called the *penetration depth*, which is the distance in the material that the evanescent wave travels before the wave's amplitude drops to $1/e$ of its initial value. Because the evanescent wave decays exponentially, the absorption coefficient is in the *argument* of the natural exponent (that is, the value that you are taking the natural exponent of). The electric field for the evanescent wave looks like the following:

$$E_{\text{trans}} = E_t e^{-\alpha z}$$

The important feature of this equation is to show the exponential decay in the amplitude of the evanescent wave. The penetration depth is chosen as that distance where the argument of the natural exponent is –1, which makes z (which is labeled δ when it's the penetration depth) equal to $1/\alpha$. Using this information makes the penetration depth

$$\delta = \frac{1}{\alpha} = \frac{\lambda_0}{2\pi n_2 \sqrt{\dfrac{\sin^2\theta}{n^2} - 1}}$$

To find a different value for the transmitted electric field amplitude in the transfer material, you simply solve for z with the desired value of E_{trans}. That change gives you the distance you need to separate the two high-index materials from each other to capture the appropriate value of amplitude in the evanescent wave (such as you would in the cube beam splitter shown in Figure 10-2).

Suppose that you want to find the penetration depth for a glass-air interface with 500 nanometer light incident at an angle of 65 degrees. Using the index of refraction value of 1.50 for glass, you get a relative index of refraction of 1/1.50, which is 0.667. The penetration depth equals $1/\alpha$, so you find this value with the following equation:

$$\alpha = \frac{2\pi n_2}{\lambda_0} \sqrt{\frac{\sin^2\theta}{n^2} - 1}$$

$$\alpha = \frac{2\pi(1.00)}{0.500 \text{ micrometers}} \sqrt{\frac{\sin^2 65°}{(0.667)^2} - 1} = 11.573/\text{micrometers}$$

To then find a different amplitude of electric field — say, 50 percent — you set up the exponential term (from the E_{trans} formula) and solve for z as follows:

$$0.5 = e^{-\alpha z}$$

$$\ln(0.5) = -\alpha z$$

$$z = \ln(0.5)/(-11.573/\text{micrometer}) = 0.060 \text{ micrometers}$$

To capture the evanescent wave at 50 percent of the incident amplitude, you need to place the second high-index material 0.06 micrometers away from the first surface where total internal reflection took place. This situation transmits about 25 percent of the incident irradiance into the second high-index material. Because the decay is exponential, any small shift in distance makes a big difference in the amount of irradiance transmitted by frustrated total internal reflection.

Chapter 11

Running Optical Interference: Not Always a Bad Thing

In This Chapter

▶ Understanding optical interference

▶ Checking out devices that utilize optical interference

*I*n daily life, interference is often something people want to avoid. You don't want someone to interfere with the work you're doing, and in communications, you don't want interference to mess up your signal. But in optics, interference isn't necessarily bad. *Optical interference* is simply the interaction of two or more light waves.

Many applications, such as medical imaging, quality control instrumentation, and holograms, use this particular property of light to make life better or more fun. Because optical interference is useful in so many applications, you need to understand the basics of this phenomenon, especially what makes optical interference visible.

In this chapter, I define terms involved with optical interference and present the conditions necessary to observe the results of optical interference. I examine the differences between the two types of optical interference and discuss some practical arrangements that use the particular types of interference.

Describing Optical Interference

Wave interference isn't unique to mechanical waves such as water waves or sound waves. Any wave phenomenon can interact or interfere with any other similar type of wave. Under the proper conditions, you can see the interference of the light waves, similar to what you see with two water waves interfering. (I go over some of the basic ideas about interference of mechanical waves in Chapter 2.) In this section, I describe constructive and destructive interference in light waves, and I explain the conditions an interference situation must meet for you to see that interference.

On the fringe: Looking at constructive and destructive interference

Constructive and destructive interference represent the two extreme situations in wave interference. The type of interference depends on how the crests and troughs of the waves overlap:

- ✔ **Constructive interference:** The crests of both waves overlap, and the resulting height is the sum of the crests of the two waves.

- ✔ **Destructive interference:** The crest of one wave overlaps the trough of the other wave, so the waves cancel each other out.

For other values of phase — that is, different alignments of crests to troughs of the waves — the interference is in between these two values.

You can envision mechanical wave interference by looking at a smooth pool of water. If you drop one stone into the pool, you see the wavefronts of the resulting water waves, which are the expanding circles that originate from where the rock entered the water. If you drop two stones separated by a certain distance (not right next to each other), you can see the result of the water waves made by each stone. Some areas appear flat, and other areas appear higher than the single water waves alone, depending on whether the interference is destructive or constructive.

In optics, fringes result from interference of light waves. The *fringes* are the areas where constructive and destructive interference occur. A *bright fringe* is one that results from constructive interference between two or more light waves. A *dark fringe* is one that results from destructive interference between two or more light waves; they appear black because the light waves cancel each other out.

In most applications in optics, the fringe pattern that optical interference produces is what makes the interference useful. However, fringes aren't readily apparent, even though many light sources surround you. Some special conditions must be met in order for you to see light fringes. In Chapter 5, I discuss how restricting the number of rays hitting a screen helps clarify an image by reducing offset images. Unfortunately, this method isn't enough to be able to see optical interference. Although offset fringes produced by different sets of light waves can blur out an individual pattern, three more conditions must be met before you can see optical interference — see the following section.

Noting the conditions required to see optical interference

Light waves are different from mechanical waves, the biggest difference being that the wavelength of visible light is much smaller than typical mechanical waves (even sound waves). This difference means that you need special arrangements to be able to see the results of optical interference. For example, you don't see interference fringes in the light reflecting off this page, so something more must be necessary to cause visible optical interference. This section covers the three special conditions that must be present for you to see fringes produced by optical interference.

Light waves must have the same wavelength

For you to see the effects of optical interference, the light waves must have the same wavelength so that you can see a constant fringe pattern. If waves with different wavelengths (or frequency) interfere, you detect the phenomenon of beats.

Beats are the periodic change in the loudness of the sound when the sound waves of different frequency interfere. Musicians use this idea for tuning instruments. One note is the *reference,* the one everyone is trying to sound like, and the other note is sounded by the instrument being tuned. The closer the note gets to the reference frequency, the slower the change in loudness. This change also happens with light waves, but you can't see the beat frequencies with your eyes (although it's still technically interference). You need a very fast detector to detect the beats, but it's possible.

Light waves must have a component of their polarizations parallel to each other

For you to be able to see optical interference, the light waves must have a component of their electric fields parallel to each other, and electric field vectors must overlap so that you can add the vectors together. If two light waves are linearly polarized, but one is vertically oriented and the other horizontally polarized, no parts of their electric fields overlap, so you don't get any interference; you have nothing to add together. In order to most clearly see this effect, the light waves must have the same polarization parallel to each other.

Light waves must be coherent

No, light waves don't need to be able to construct logical arguments to interfere. *Coherence* is a physics term that describes how the crests or troughs of all the waves line up. For interference to occur, the crests don't have to line up exactly. In fact, the particular phase difference between waves isn't important; coherence looks at how long the particular phase relationships last.

For you to be able to see light waves interfering, the phase relationship between the waves must be fixed for a long enough period of time so that you can see the fringe patterns. Light waves from light bulbs aren't coherent because the phase relationship between the waves changes randomly every 10^{-14} seconds or so. That's why you don't see interference fringes dancing around a room lit up by light bulbs; the fringes don't stick around long enough, even if you could isolate one set of fringes from the myriad number of randomly oriented fringes that appear in the same general area.

You can use a couple of tricks to make light coherent. In this day and age, an easy option is to use a laser. All you have to do is turn on a laser, and you have a bright source of coherent light. (In Chapter 15, I give you an idea why lasers produce coherent light.) Because lasers are so readily available, they're the go-to device for using optical interference as a measuring tool.

The second trick is to use a small slit or pinhole in a sheet of metal or thick piece of paper. The hole creates a *point source* of light (a light source that sends light out from a single point rather than many points), and the light from this pinhole generally has enough coherence to see fringes. Like the pinhole camera I discuss in Chapter 5, the bright fringes aren't easily observable because of the very small amount of light getting through the pinhole. Although this method was used in the first experiment to measure the wavelength of light, it's not as handy for optical interference as using a laser is.

Perusing Practical Interference Devices: Interferometers

Interference requires two or more waves, so special arrangements to bring several waves together with the proper conditions are necessary for you to be able to see the effects. You can use arrangements called *interferometers* to see interference for particular applications. The two categories of interferometers are wavefront-splitting interferometers and amplitude-splitting interferometers, and I discuss them in the following sections.

Wavefront-splitting interferometers

One basic arrangement utilizes interference by splitting the wavefront: the creatively named *wavefront-splitting interferometer*. It's important from a historical perspective because it was the first arrangement that allowed Thomas Young to determine the wavelength of light. In a pedagogical sense, it's a relatively simple experiment to set up so that students can see optical interference for themselves and make measurements using the interferometer.

A wavefront-splitting interferometer splits a single wavefront in some fashion and brings each part together to interfere on a screen. You can find a few experimental arrangements that involve wavefront splitting, but they're variations of the two-slit arrangement. The following sections dive into some of these interferometers.

The two-slit arrangement

In a two-slit interferometer, coherent light is incident on an opaque sheet with two narrow slits carved in it. Figure 11-1 illustrates the general arrangement for this situation.

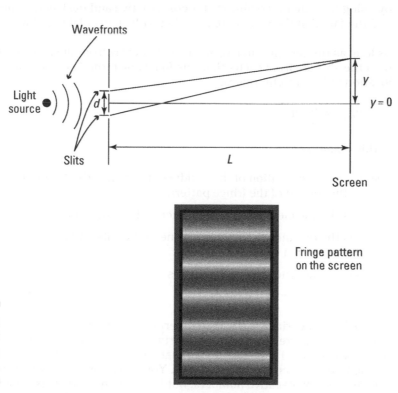

Figure 11-1:
Two-slit interference in action.

One part of the wavefront goes through one slit, and another part of the same wavefront goes through the other slit. Although diffraction causes the light to expand after passing through the slits, I want to talk about the closely spaced fringes that result from interference between the two different parts of the incident wavefront.

A varying light pattern appears on the screen because the light travels different distances from the respective slit to the screen. Because the path difference between the waves varies with the point of the screen, the light pattern changes from a bright fringe to a dark fringe and back and forth until the irradiance is so low that you can't see it with your eye anymore. (Flip to the earlier section "On the fringe: Looking at constructive and destructive interference" for more on fringes.)

Figure 11-1 shows a special point, indicated by $y = 0$, right in the middle of the screen that's directly opposite the midpoint between the two slits. This point serves as the reference point to measure the centers of the other fringes, both bright and dark. You can find the location of the bright and dark fringes by using the general conditions for constructive and destructive interference and the fact that the waves are modeled with sinusoidal functions.

As long as the screen is much, much farther away from the plane of the slits than the slits are from each other, the following formula lets you find the locations of the bright fringes:

$$y_{bright} = \frac{\lambda L}{d} m$$

In this equation,

- y_{bright} is the location of the middle of the mth order bright fringe relative to the middle of the fringe pattern, $y = 0$.
- λ is the wavelength of the light incident on the slits.
- L is the distance between the plane of the slits and the screen where the fringe pattern appears.
- d is the distance between the slits.
- m is the order of the fringe.

The *order* of the fringe is a way to determine the number of fringes from the very center. The zeroth-order bright fringe appears at $y = 0$. The first-order bright fringe ($m = 1$) appears above that, and the higher-order fringes, labeled by the integer m, count up from there. You can refer to the fringes below the zeroth order by negative numbers, but because the pattern is symmetric around $y = 0$, the distinction of positive and negative orders isn't important.

Under the same conditions used with finding the bright fringes, you can locate the dark fringes with the following formula, which uses the same variables as the bright-fringe formula:

$$y_{dark} = \frac{\lambda L}{d}\left(m + \frac{1}{2}\right)$$

A bright fringe occurs in the exact middle of the fringe pattern (because the path length is exactly the same distance from each slit), so two zeroth-order dark fringes appear, one on either side of the zeroth-order bright fringe. Because the positive and negative orders have no practical difference, you can work with either dark fringe. The y value still remains the distance from the center of the fringe pattern, $y = 0$, to the center of the dark fringe.

In a laboratory situation, you may be given a two-slit slide and asked to determine the wavelength of a particular laser. With the spacing between the slits known, you can easily measure the distance from the screen to the plane of the slits, find the center of the fringe pattern, and measure the distances to the centers of several bright and dark fringes (measure different y_{bright} and y_{dark} values). Using this data with the bright/dark fringe equations, you can solve for the remaining unknown, which is the wavelength, just like Thomas Young did in 1801.

When you actually set up the two-slit arrangement, you may notice that the fringes aren't all the same brightness. Depending on the size of the slits, the zeroth-order bright fringe is the brightest, and the other orders are dimmer. The reason for this effect is diffraction, which I discuss in Chapter 12. For some slit arrangements, the intensity is nearly the same for all the bright fringes near the center. Don't be surprised if you see either of these situations. Even though diffraction and interference happen simultaneously, the two don't distort the measurements that I present in this section. The equations here deal with the spacing of the fringes, not the brightness.

Fresnel's bi-prism

A *bi-prism arrangement* produces a fringe pattern identical to the fringe pattern produced by the two-slit arrangement but does away with the slits by splitting the wavefront with refraction instead. You can see a basic bi-prism arrangement in Figure 11-2.

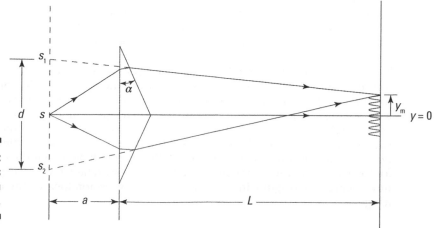

Figure 11-2:
Fresnel's
bi-prism
arrangement.

The source, S, in Figure 11-2 is a point source, which can be made with a pinhole or a slit that's placed a distance of a from the apex of the bi-prism. Refraction splits the wavefront by bending the portion of the wavefront that hits the lower part of the prism upward and the portion that hits the upper part of the prism downward. Because the interference pattern is identical to the two-slit pattern, you can use the fringe equations in the preceding section. The only thing you need to calculate is the number to use for the separation of the virtual sources S_1 and S_2 (which is the slit separation, d, for the two-slit situation) created by refraction of the two parts of the wavefront. The expression for this distance is given by

$$d = 2a(n-1)\alpha$$

In this equation,

- d is the separation distance between the virtual sources.

- a is the distance the source of light is from the base of the prism.

- n is the index of refraction of the prism material.

- α is the prism angle, measured in radians.

Using this information, you can measure the fringe spacing and determine the wavelength of the source similar to what you do with the two-slit arrangement. The advantage of the Fresnel's bi-prism is that it's a little less complicated than the two-slit arrangement. Some commercial devices called *wavelength meters* use this concept to measure the wavelength of light sources.

Lloyd's mirror

A second variation of the two-slit arrangement is called *Lloyd's mirror*. This arrangement is the simplest variation of the two-slit arrangement (though the fringe analysis is more complicated) and is shown in Figure 11-3. The Lloyd's mirror arrangement is also the most versatile wavefront-splitting interferometer, in that it has been used for a large number of wavelengths in the electromagnetic spectrum from x-rays with crystal surfaces to radio waves from the surfaces of lakes or even the ionosphere.

The fringe pattern produced by Lloyd's mirror is actually *complementary* to the pattern produced by the two-slit arrangement. That is, the top of the mirror, which is equivalent to the middle of the fringe pattern, is actually a dark fringe. As I mention in Chapter 10, when light bounces off a surface with a higher index of refraction, the reflected wave is shifted by half of the wavelength of the light. This phase shift in the reflected wave (which doesn't occur in the light that travels directly to the screen) means that the equations for the location of the bright and dark fringes are switched; the y_{dark} expression now locates the bright fringes and the y_{bright} expression locates the dark fringes.

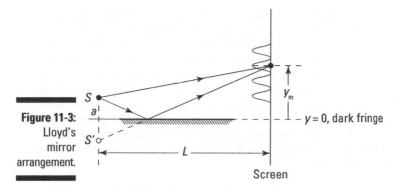

Figure 11-3:
Lloyd's
mirror
arrangement.

Another difference is that the mirror blocks the lower half of the screen so that only one set of fringes is viewable above the mirror. Instead of having positive and negative values for the order number *m*, you only see fringes with positive numbers. The value you use for *d* is twice the distance that the source is above the surface of the mirror (the distance *a* in Figure 11-3).

Amplitude-splitting interferometers

Amplitude-splitting interferometers involve beams of light rather than large wavefronts. Optical devices called *beam splitters* separate part of the beam and send it in a different direction. These beam splitters can be made with a *partially silvered* mirror (one that reflects some of the light and transmits part of the light) or a cube beam splitter (such as the one I describe in Chapter 10). The following sections describe several amplitude-splitting interferometers.

The Michelson interferometer

The *Michelson interferometer* involves the interference of light in two separate arms. Check out a basic Michelson interferometer in Figure 11-4.

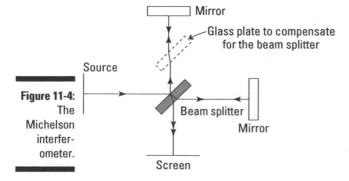

Figure 11-4:
The
Michelson
interfer-
ometer.

A Michelson interferometer sends laser light into a beam splitter that splits the laser into beams of equal irradiance. Each beam reflects off a mirror, which directs the beam back to the beam splitter. In the beam splitter, the two beams recombine and go out to a screen or a camera where you can see the fringe pattern. If the beams have a small amount of divergence, the fringe pattern looks like a bull's-eye pattern with concentric bright and dark rings. More often, though, the fringes are vertical or horizontal lines. If you move one of the mirrors by a small amount, the fringe pattern shifts proportionally to the distance that the mirror moved, provided the other mirror doesn't move. This feature allows a Michelson interferometer to make very precise measurements of distances.

Distance measurements rely on the fact that to make a dark fringe change to a bright fringe or vice versa, the phase needs to shift by π, corresponding to a path length difference shift of $\lambda/2$. So if you see a fringe change from bright to dark or dark to bright, you know that one of the mirrors in the interferometer moved a distance equal to $\lambda/4$ because the light has to make to two trips over the same distance (from the beam splitter to the mirror and back to the beam splitter). Because the wavelength of light is on the order of 500 nanometers, the Michelson interferometer can measure distances that are as small as 125 nanometers.

If the fringe changes many times and you keep track of the number of changes from bright to dark to bright, you can multiply the number of changes by $\lambda/2$ to get the distance that the mirror has moved (for distances larger than the wavelength of the light).

The Twyman-Green interferometer

The Twyman-Green interferometer uses light from a point source (rather than an extended source like the Michelson does) and a *collimating lens,* which makes all the waves travel parallel to each other, thus creating a *collimated beam.* Figure 11-5 shows this variation of the Michelson interferometer.

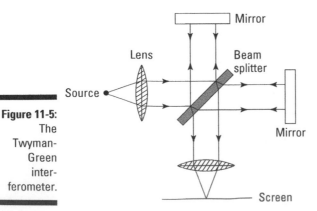

Figure 11-5:
The Twyman-Green interferometer.

After the beam recombines, it's focused at the screen or detector. This arrangement creates a series of equally spaced interference lines. If you place an optical element in one of the arms, any surface or internal defects in the optics shows up as distortion in the fringe pattern. This interferometer is a very important quality control arrangement for manufacturers of optical systems. It allows them to select optics that have a minimal number of defects.

The Mach-Zehnder interferometer

The _Mach-Zehnder interferometer_ arrangement separates the paths so that the light doesn't travel along the same path twice. Figure 11-6 illustrates this expanded form of the Michelson interferometer.

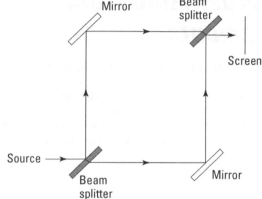

Figure 11-6:
The Mach-Zehnder interferometer.

The nonrepeated path allows you to measure small changes pressure or thermal waves in air create in index of refraction. Mach-Zehnder interferometers have been used by aerospace companies to study airflow and pressure changes for airfoils and aircraft bodies. The Mach-Zehnder interferometer is an important arrangement for placing optical information onto a laser beam. I present this topic in more detail in Chapter 21.

The Sagnac interferometer

The _Sagnac interferometer_ is different from the mirrored amplitude-splitting interferometers mentioned earlier. The beam is split, and each beam travels around the same path but in opposite directions. The beam's traveling in opposite directions means that it has no fringe output; you can't see a fringe pattern. You can see a typical Sagnac arrangement in Figure 11-7.

If you rotate the interferometer about an axis that's perpendicular to the plane of the interferometer, the frequency of the light changes proportional to the velocity that the mirror is moving relative to the incident light as the light travels relative to the moving mirrors. This frequency shift is proportional to the rotation speed and can be detected by a high-speed photodetector. This arrangement is used in optical gyroscopes to help keep satellites in the proper orientation.

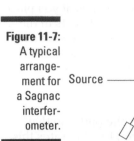

Figure 11-7:
A typical arrangement for a Sagnac interferometer.

Accounting for Other Amplitude-Splitting Arrangements

Interferometers aren't the only situations that use amplitude splitting to create optical interference. Another important situation involves the use of *thin films* — either thin coatings of insulating material placed (usually) on a glass substrate or just simply a very small thickness of air between two surfaces. Two useful applications of this concept include an important optical technology of thin dielectric film coatings and quality control or small detail measurement.

Thin film interference

Thin film interference is the arrangement of interference between the light reflected from the top and the bottom of a thin dielectric coating. (And by *thin*, I mean something on the order of 10^{-6} meters or smaller.) This phenomenon can also take place with light that travels through the material (transmitted light as opposed to reflected light), but here I concentrate on the simpler problem of dealing with reflected light to give you an idea about how this works. Thin film interference is a very important interference phenomenon in optics because it provides a very controllable way to design how much of the incident light is reflected at a particular wavelength. It can also eliminate reflected light on lenses (and other optical components such as windows) to help improve the operational efficiency of the system. Anti-reflection thin film coatings on eyeglasses reduce the reflection of light from your glasses (to minimize the appearance of the flash in your eyes when someone takes your picture).

When designing a thin film coating to produce either constructive or destructive interference, keep the following in mind: Any time light encounters a different medium, some of the light is reflected, and the rest is transmitted. In the situation of external reflection in Chapter 10, the reflection coefficient shows a phase change of $\lambda/2$ in the reflected wave. You must take this change into account when determining the final phase relationship between the reflected waves.

If the film is thin enough that it doesn't absorb a significant amount of the light, light reflects off the bottom of the film and heads back through the top, where it can interfere with the light that reflected off the top of the film. The film thickness can shift the phase difference between the two waves such that they would interfere *destructively*, canceling each other out (anti-reflection coating), or *constructively*, adding to each other (mirror coating). Figure 11-8 shows a typical arrangement for thin film interference.

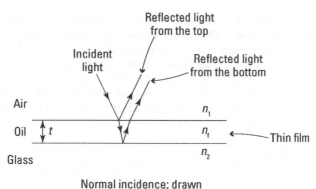

Figure 11-8:
Thin film
interference.

To calculate the film thickness, t, necessary for constructive or destructive thin film interference, just follow these steps:

1. **Identify all the reflected rays in the situation.**

2. **Calculate the total phase change produced by reflections that happen as the light goes from a low index of refraction material to a high index of refraction material ($\lambda/2$ shift each time the ray reflects in such a manner).**

3. **Determine whether you need to preserve the phase shift produced by the reflections or shift it by $\lambda/2$, depending on whether you need constructive or destructive interference.**

   If you need to shift the phase difference, set $2n_f t = (m + 1/2)\lambda$. If you need to preserve the phase difference, set $2n_f t = m\lambda$.

In these equations,

- n_f is the index of refraction of the thin film.
- t is the film thickness.
- m is the order number.
- λ is the wavelength of the incident light.

Suppose you want to design an anti-reflection coating for eyeglasses. The magnesium oxide material has an index of refraction of 1.37, and the polycarbonate material that the lenses are made of has an index of refraction of 1.52. You want the coating placed on the lenses so that it cancels out light with a wavelength of 500 nanometers. You want to find the thinnest coating that will work. For simplicity's sake, you look only at the case where the light is normally incident on the lenses.

To identify the reflected rays, assume that the lenses are thick enough that they don't reflect light of any significance from the back surface of the lens, so you can neglect that reflection. You only need to account for the rays that reflect from the front surface, the surface with the coating. You find two rays, one from the top of the magnesium oxide film and one from the top surface of the lens at the bottom of the thin film. For any rays that reflect off a higher-index material, the reflected wave is shifted by $\lambda/2$. If the light reflects off a lower-index material, no shift exists.

Here's a rhyme that may help you remember when to add the phase shift to the reflected ray: "Low to high, shift by pi." Pi is the shift in phase angle, which corresponds to a path length shift of $\lambda/2$, and pi rhymes better.

In the anti-reflection coating, both reflected rays bounce off a higher-index material. Therefore, each ray is shifted by $\lambda/2$, so they're *in phase* when they get above the thin film because they have zero phase difference due to reflection.

Because you're dealing with an anti-reflection coating, you want destructive interference so that the reflected light cancels out and you can't see it. A phase difference of zero between the two reflected waves results in constructive interference. You need to shift the phase between the two waves by making the path difference of the ray that travels through the thin film correspond to an odd integral multiple of $\lambda/2$. Here's the math:

$$2n_f t = (m + 1/2)\lambda$$
$$2(1.37)t = (0 + 1/2)500 \text{ nm}$$

You use the zeroth order because you want the thinnest film that will cause destructive interference, but you can't have a film thickness of zero. This calculation gives you a thickness of 91.2 nanometers. Notice also that you have an anti-reflection coating for other m values, so thicknesses of 274 nanometers, 456 nanometers, and so on also technically work. As a matter of practice, though, the thicker you make the thin film, the worse the cancelation, so the thinnest film possible always works best.

Newton's rings

Newton's rings is the name of the interference pattern created by putting a curved optic, such as a convex lens, on a very flat glass surface. The flat glass keeps the irradiances at nearly the same level; a mirror usually doesn't work well because the two irradiances are so different that the fringes aren't visible. This arrangement (shown in Figure 11-9) is a special case of thin film interference where the air gap is the thin film and its thickness varies from the center of the lens.

The interference fringes appear between reflections from the curved surface of the lens closest to the glass and the top surface of the glass closest to the lens. Because most lenses are spherical element lenses, the fringe pattern appears as a series of concentric rings.

The very center of the fringe pattern is always dark because the center of the lens is in contact with the flat glass, but you still have an infinitesimal air gap between the two surfaces. Therefore, the light reflected in the lens shifts by $\lambda/2$ relative to the light reflected off the glass so that the two waves interfere destructively. Then the different air gap thicknesses alternatively vary between λ and $\lambda/2$, which makes a dark or bright fringe.

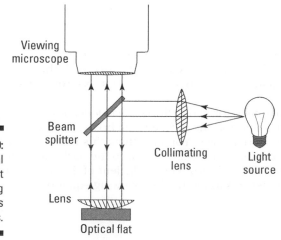

Figure 11-9:
A typical arrangement producing Newton's rings.

Viewing microscope

Beam splitter

Collimating lens

Light source

Lens

Optical flat

The Newton's rings arrangement is often a simpler arrangement for evaluating the quality of lenses than the Twyman-Green interferometer is (see the earlier section), but any imperfections in the optics show up as a distortion in the fringe pattern as they do with the Twyman-Green.

Fabry-Perot interferometer

An interferometer device that uses multiple reflections to produce a transmitted interference pattern (not just two reflected beams) is called a *Fabry-Perot interferometer*. A typical arrangement is shown in Figure 11-10. *Note:* I classify this interferometer as an "other amplitude-splitting arrangement" because the analysis is quite different from the other amplitude-splitting interferometers.

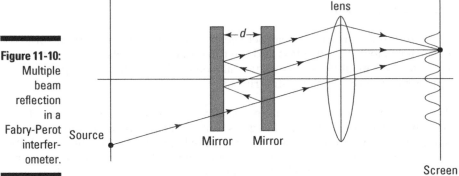

Figure 11-10: Multiple beam reflection in a Fabry-Perot interferometer.

Although this interferometer can be made with flat pieces of glass placed parallel to each other, the device is much more effective when you use high-reflectivity mirrors.

Using Fabry-Perot interferometers relies on parameters that are based on the mirrors' coefficients of reflection, r (see Chapter 10). With this value, you can find two important characteristics of Fabry-Perot interferometers.

The first important parameter is called the *coefficient of finesse*. This quantity gives a measure of the *contrast*, the difference in brightness between the bright fringes and the dark fringes. You may think this value always seems very high, but you can't make perfect mirrors, so the dark fringes never actually reach an irradiance of zero. The coefficient of finesse, F, for a Fabry-Perot interferometer is found by the following equation:

$$F = \frac{4r^2}{\left(1 - r^2\right)^2}$$

As you use mirrors with larger reflection coefficients, the dark fringes become darker, and the bright fringes become narrower. One of the main uses of the Fabry-Perot interferometer involves spectroscopy, and being able to tell the difference between two closely spaced wavelengths is a key feature in being able to identify materials based on their spectral fingerprint.

Another important parameter of a Fabry-Perot interferometer is the finesse. (***Remember:*** This quantity is different from the coefficient of finesse, so be sure that you don't confuse the two.) The *finesse* of the interferometer is the ratio of the separation (in wavelength) between adjacent bright fringes (called the *free spectral range* of the device) to the width (in wavelength) of the bright fringes. The finesse of a Fabry-Perot interferometer is given by

$$\mathcal{F} = \frac{\pi}{2}\sqrt{F} = \frac{\pi r}{1 - r^2}$$

The finesse relates more directly to the resolving power of a Fabry-Perot interferometer. *Resolving power* is the capability of an instrument to tell the difference between two closely spaced wavelengths. If two light waves enter the Fabry-Perot interferometer with wavelengths that are closer together than the resolving power of the instrument, the instrument shows only one bright fringe. If you increase the resolving power by, say, increasing the r value, two peaks appear, corresponding to the two different wavelengths. The resolving power again shows that the mirrors' reflection coefficient determines the performance of the Fabry-Perot interferometer. The separation of the mirrors determines the free spectral range of the device, which is determined by the particular wavelengths used for an application (but that's outside the scope of this book).

Chapter 12

Diffraction: Light's Bending around Obstacles

*J*ust like wave interference (see Chapter 11), diffraction is not unique to mechanical waves like water waves or sound waves. In fact, you find little physical difference between interference and diffraction; they both add similar waves together to get the result. Generally, the analysis of interference involves two or a few waves, and diffraction involves many waves. In both situations, the result of diffraction and interference is that a fringe pattern (areas of bright and dark) appears on a screen due to constructive or destructive interference between the waves that combine at a particular point on a screen.

However, interference doesn't explain why light expands behind an obstacle placed in the way. Diffraction models the property of light waves to expand and tend to fill the area behind an obstacle such as a pinhole or a slit in an opaque screen.

In this chapter, I describe the two perspectives you can use to look at light after it passes through an obstacle. Within the realms of these two types of diffraction, I also cover some special situations that explain why you see certain things in some interference arrangements, and I introduce a couple of very important optical devices.

From Near and Far: Understanding Two Types of Diffraction

Diffraction is light's response to having something mess with its path, so diffraction occurs only when something blocks part of the wavefront. A lens, even though it's transparent, has edges that work just like an obstacle. If you take away the obstacle, you don't get any diffraction because nothing disturbs the light.

The two types of diffraction aren't due to different phenomena; they're distinct because of where you're looking relative to the obstacle. This difference has major implications in imaging applications, from telescopes looking at objects far away to photolithography used in semiconductor device fabrication (where the image may be very close to the lens). Here, I discuss the two types of diffraction and help you recognize which kind you're dealing with.

Defining the types of diffraction

The type of diffraction depends on whether the light you're observing is near or far from the *aperture* (opening through which light passes). Here are the two types of diffraction:

- **Fresnel diffraction:** *Fresnel diffraction* looks at what happens to light very near the aperture, so it's often called *near-field diffraction*. For you to see this type of diffraction, you place an opaque screen with a pinhole or a narrow slit in a beam of light and then place another screen very close to the pinhole. Generally, you can clearly see light that looks exactly like the pinhole, except the edges appear a little bit fuzzy. If you slowly move the screen away from the pinhole, you can still tell the shape of the pinhole from the light on the screen, but now you start to see fringes appear in the image of the pinhole. As you move the screen farther away, the fringe pattern changes, and the overall image increases in size.

- **Fraunhofer diffraction:** *Fraunhofer diffraction* looks at what happens to light very far from the aperture. To see this type of diffraction with a slit or pinhole obstacle, you place the screen far away from the aperture. In this setup, the light pattern that you see doesn't look anything like the slit or pinhole that caused the light to diffract. And in contrast to Fresnel diffraction, the pattern remains the same (the bright and dark areas don't change relative location) as you move the screen farther away from the aperture; the overall size of the pattern simply increases in size.

Determining which type of diffraction you see

Although the light patterns produced by the different types of diffraction are of different quality, figuring out which type you see is hard to tell by just looking at the light pattern. However, you can evaluate a couple of physics factors to come up with a quantitative way to distinguish what type you're looking at:

✔ **Shape of the incident wavefronts:** If either the wavefronts incident on the aperture or the wavefronts from the aperture incident on the screen are curved, the diffraction is Fresnel type. If the wavefronts are planar on the aperture or the wavefronts from the aperture incident on the screen are planar, the diffraction is Fraunhofer type.

✔ **Relative strength and phase relationship of the light waves that pass through the aperture:** Conceptually, you can model the diffraction pattern by dividing the aperture into a large number of very small sections and tracing the electric field from each section to a point on a screen. With Fresnel diffraction, where the screen is close to the aperture, the electric fields have nearly the same amplitude, but the path difference from each section to the screen can be significant. The resulting intricate fringe pattern comes from the addition of all the waves incident at the point on the screen; the fringes are very closely spaced and generally reflect the shape of the aperture.

With Fraunhofer diffraction, as you move the screen far away from the aperture, the path difference between the light waves from the different sections becomes less important because it's very small compared to the distance traveled. The electric field amplitudes from the waves stay more or less the same, but the angle that their paths make with the line perpendicular to the screen is what determines the fringe pattern. The Fraunhofer diffraction patterns are typically much simpler than the Fresnel diffraction patterns, as I show you later in the chapter.

These conditions are still not very easy to use to determine what type of diffraction you see. Several mathematicians found the following relatively simple equation to determine what kind of interference you see:

$$L > \frac{a^2}{\lambda}$$

In this equation,

✔ L is the smaller of the two distances from the source to the aperture and aperture to the screen.

> ✔ *a* is the greatest width of the aperture (diameter for a pinhole and width or height for a rectangular hole).
>
> ✔ λ is the wavelength of the light incident on the aperture.

If the smallest distance is in fact greater than a^2/λ, the diffraction is Fraunhofer. If the smallest distance doesn't satisfy this condition, the diffraction is Fresnel. This condition provides a nice, quantifiable way to determine which type of diffraction you see; it's much easier to work with than using the factors I mention earlier, but the list of factors is helpful for remembering the characteristics of the diffraction patterns that you see in each category.

A distinguishing feature of Fraunhofer diffraction is that the central bright fringe is larger than any other fringes in the pattern. This characteristic is another way you can generally tell what phenomenon is responsible for the fringes. If the fringes aren't equally spaced, they're the result of diffraction. If the fringes are equally spaced, they're due to interference of light from more than one source.

Going the Distance: Special Cases of Fraunhofer Diffraction

Fraunhofer diffraction is observed more frequently than Fresnel diffraction and plays a very significant role in imaging applications. Because most optical devices operate in the far-field region, Fraunhofer diffraction is much more common than the Fresnel variety.

A significant feature of Fraunhofer diffraction is that the diffraction pattern you see doesn't look like the aperture that the light passed through. Usually, many more bright fringes appear around the central region that aren't present in the actual aperture.

In Fraunhofer diffraction, the screen is typically very far from the obstacle, so the light waves have all traveled approximately the same distance no matter what part of the aperture they went through. At any point on the screen, the contribution coming from each part of the aperture has the same amplitude and relatively little phase variation, so the fringe patterns tend to be simpler. I show some of the special cases of Fraunhofer diffraction in the following sections.

Fraunhofer diffraction from a circular aperture

Analyzing light passing through a circular aperture involves dividing the aperture into small, infinitesimal regions, propagating the light from each of these regions to a screen, and adding up the waves. This process can become rather complicated, but a special mathematical function called a *Bessel function of the first kind* turns out to be a helpful solution. The only thing you need to concern yourself with is what this function generally looks like. Figure 12-1 shows an example of Fraunhofer diffraction from a circular aperture.

Figure 12-1:
Fraunhofer
diffraction
from a
circular
aperture.

The bright disk in the center of the diffraction pattern is called the *Airy disk*. The size of this disk determines an important aspect of imaging applications called resolution. *Resolution* is a measure of how close two objects can be while you can still tell that there are two objects.

In order to determine the resolution, you need to be able to calculate the size of the Airy disk. A convenient (and practical) way to determine the size is to measure the radius of the disk from its center to the middle of the first dark ring, where the irradiance drops to zero. In general terms, you can more conveniently talk about the size in terms of the far-field angle (the angle formed between a line from the observation point to the center of the disk and a line from the same point to the middle of the first dark ring) given by

$$\theta_{\min} = \frac{1.22\lambda}{D}$$

In this equation,

- θ_{min} is the far-field angle of the Airy disk.
- λ is the wavelength of the light incident on the aperture.
- D is the diameter of the circular aperture.

The 1.22 factor in this equation comes from the Bessel function and is a consequence of the aperture being a circle.

Using the Fraunhofer diffraction pattern from a circular aperture, the *Rayleigh criterion* provides a practical definition of resolution. This criterion states that if the central maximum of one image falls on the first minimum of another image, the images are said to be *just resolved*. You can see the situation of resolved images to unresolved images in Figure 12-2.

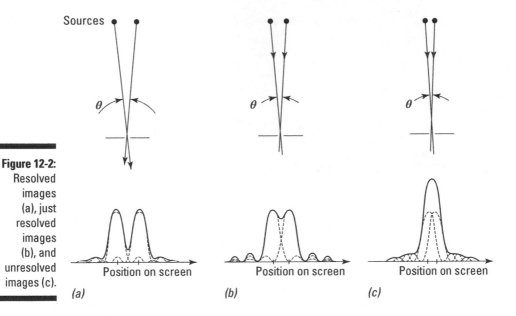

Figure 12-2: Resolved images (a), just resolved images (b), and unresolved images (c).

Using the equation to find the size of the Airy disk, the equation for the minimum angular separation between two objects that can be just resolved is

$$\theta_{min} = \frac{1.22\lambda}{D}$$

Although the angle is generally the most useful, you can use the arc length relation to find the distance separating two objects that can be just resolved

$$d = L\theta_{min}$$

In this equation,

- ✔ d is the distance between two objects.
- ✔ L is the distance from the plane that contains the objects to where the aperture is located.
- ✔ θ_{min} is the minimum resolvable angle of separation between two objects.

Suppose you want to determine the closest to each other that two objects can be on the moon such that you can still resolve them with your telescope, which has an aperture of 15 centimeters. The moon is approximately 400,000 kilometers away from the earth, and you can use 600 nanometers for the wavelength of the light that you're using. First, you find the minimum half angle:

$$\theta_{min} = \frac{1.22\lambda}{D}$$

$$\theta_{min} = \frac{1.22\left(600 \times 10^{-9} \text{ meters}\right)}{15 \times 10^{-2} \text{ meters}}$$

This equation gives you an angle of 4.88×10^{-6} radians. To convert this result to a distance between objects on the moon, you use the arc length equation:

$$d = L\theta_{min}$$

$$d = \left(4.00 \times 10^{8} \text{ meters}\right)\left(4.88 \times 10^{-6} \text{ radians}\right)$$

Your distance is 1,950 meters. If two objects on the moon are closer together than this distance, you won't be able to distinguish with your telescope that two objects exist at that point on the moon.

Fraunhofer diffraction from slits

The two-slit experiment I introduce in Chapter 11 is less complicated than the circular aperture, and it leads to a very useful device in spectroscopy: the diffraction grating. To understand how a diffraction grating works, I show you how a single slit works and then look at what happens as you add more slits.

Fraunhofer diffraction from a single slit

If you place a screen a very large distance from an opaque screen with a narrow slit in it, you see a pattern that isn't simply an image of the slit but rather bright and dark fringes. The physics analysis involves dividing up the slit into a large number of infinitesimal points, propagating waves from each of these points, and interfering them at a point on the screen. The result is a pattern of fringes similar to the one in Figure 12-3.

Figure 12-3: Fraunhofer diffraction from a single slit.

With a single slit, analyzing all the waves coming from the slit gives a relatively simple equation for locating the center of the dark fringes:

$$y_{dark} = \frac{m\lambda L}{a}$$

In this equation,

- y_{dark} is the distance from the center of the central bright fringe that the mth-order dark fringe lies.
- m is the order number of the dark fringe ($m = \pm1, \pm2, \pm3$, and so on).
- λ is the wavelength of the light incident on the slit.
- L is the distance from the plane of the slit to the screen.
- a is the width of the slit.

Figure 12-4 shows the geometry for a single-slit Fraunhofer diffraction.

Remember that if you want to use this equation, the distance between the slit and the screen must be large: L must be much larger than a, which is usually pretty easy to do in a laboratory. But it's sometimes easy to overlook in tabletop arrangements.

Fraunhofer diffraction from two slits

In Chapter 11, I talk about interference between the light that travels from two separate slits. What may not be clear in the pattern is that the bright fringes aren't all the same brightness. The variation in brightness is the result of diffraction.

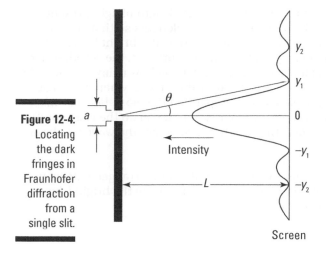

Figure 12-4:
Locating the dark fringes in Fraunhofer diffraction from a single slit.

What you see in the two-slit arrangement is the superposition of two diffraction patterns, one from each slit. Because the slits are offset from each other, the diffraction patterns at the screen are also offset, which produces interference effects between the light from the two slits. The diffraction patterns change only the brightness of the bright fringes, not the location of the bright or dark fringes produced by interference. Figure 12-5 illustrates a sample two-slit interference pattern combined with the single-slit diffraction pattern.

Figure 12-5:
Two-slit interference with a single-slit Fraunhofer diffraction envelope.

Fraunhofer diffraction from many slits: The diffraction grating

As you increase the number of slits, the bright fringes become narrower and the dark fringes, although less dark, become broader. Carried to an extreme where you have many thousands of slits placed close to each other, you get a *diffraction grating.* A diffraction grating is a very important device in *spectroscopy,* looking at the light that comes from materials to identify what the material is (see the sidebar "Separating white light").

You usually use a grating with more than one wavelength of light present. The purpose of the grating is to separate the wavelengths so that you can tell which ones are present. The location of the zeroth-order bright fringe is the same for all wavelengths. If many wavelengths are present, the zeroth order doesn't separate them and looks white, depending on how many wavelengths are present. At the highest order generated by the fringes, where the angle θ is near 90 degrees, the wavelengths are separated the most. Because most of the energy is located in the zeroth order (where it can do the least good), the energy in the highest-order fringes is the least where the wavelengths are most easily identified.

Using the same analysis you use with all the diffraction arrangements, you can find the following relatively simple equation to locate the bright fringes produced by a diffraction grating:

$$d \sin\theta = m\lambda$$

In this equation,

- d is the separation between adjacent slits.
- θ is the angle between a line perpendicular to the plane of the slits and the location of the mth-order bright fringe.
- m is the order number of the bright fringe ($m = 0, \pm1, \pm2, \pm3$, and so on).
- λ is the wavelength of the light incident on the grating.

Separating white light

If you've ever looked at a CD, DVD, or even a Blu-ray Disc with the bottom side held up to the light and seen a rainbow of colors on the surface, you've seen the effect of a diffraction grating with a many-wavelength (several-color) light source. The grooves that contain the digital data on the disc are close enough that when you hold it so that the incident light has a large incident angle, diffraction off the grooves causes the colors to separate out. The light used in common household lighting has many different wavelengths, so it appears more or less white, and according to the diffraction equation, each wavelength (color) has its own unique angle.

The separation of white light into its colors (like sunlight in small water droplets makes a rainbow) is what a diffraction grating does. This separation is also how the grating is used to identify the light from materials like liquids and gasses. Many materials have a unique optical fingerprint. Diffraction gratings are one tool that you can use to try to separate out the colors emitted to see this fingerprint and identify the material.

Many times, you find that you have the number of lines per linear distance (such as 10,000 lines per inch) rather than the grating spacing you need for the equation. To get the slit separation distance, d, that you need to use for the grating equation, you invert the number of lines. For example, for a grating with 10,000 lines per inch, the d value is 1/10,000, or 1.0×10^{-4} inches.

Blazed gratings try to change the situation where most of the light is sent into the zeroth order, where no separation exists. A *blazed grating* has the slits angled so that most of the energy is taken from the zeroth order to a higher-order fringe, where the wavelengths are more broadly spread out. Figure 12-6 shows examples of an unblazed and blazed grating. The blazed grating (on the right) has a blaze angle of +30 degrees which sends more of the energy into the m = +1 diffraction order, where it is more useful than in the m = 0 order. The amount of energy in each diffracted order is represented by a gray scale: the black line has the most energy and the light gray line has the least energy.

Figure 12-6: A schematic of an unblazed grating (left) and a blazed grating (right).

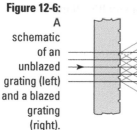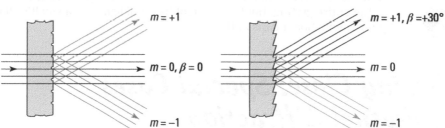

A blazed grating is a more-efficient and effective way to identify closely spaced wavelengths present in a beam of light because it puts most of the energy in the higher-order fringes.

Another important feature of a diffraction grating is the *resolving power*, which is an indication of a grating's capability to separate out two closely spaced wavelengths. The resolving power for any spectroscopic instrument is defined as

$$R = \frac{\lambda}{(\Delta\lambda)_{min}}$$

In this equation,

- R is the resolving power.
- λ is the average wavelength of the light.
- $(\Delta\lambda)_{min}$ is the smallest difference in wavelength that can be seen.

The resolving power specifically for a diffraction grating is presented in terms that are easily found for a particular grating and is given by

$$R = mN$$

In this equation,

- R is the resolving power of the diffraction grating.
- m is the order of the bright fringe used.
- N is the number of slits that the incident light illuminates.

Generally speaking, the resolving power is used to compare different gratings. Sometimes, you need to achieve a balance with working in higher orders versus the width of the light beam you need to use.

In order to use the resolving power equation correctly, you must determine how many slits the incident light illuminates. If you just use the number of slits in the entire grating, but the incident beam of light has a smaller diameter, your calculation means nothing.

Getting Close: Special Cases of Fresnel Diffraction

Fresnel diffraction is, in general, not observed as much. Certain optical arrangements can help you see it, but it often appears as subtle effects in certain optical systems that make this diffraction type a significant effect, most often when you're looking at very small things and generating a large amount of magnification.

A helpful tool to analyze Fresnel diffraction patterns is the concept of the Fresnel zones. If you analyze light from a point source, which produces spherical wavefronts, relative to a point in front of an emitted wavefront, you can divide the wavefront into thin ring zones that are concentric to a line connecting the point source and the frontal point. The ring zones in the spherical wavefront are selected such that the distance from the forward point to a point in a zone differs by a distance of $\lambda/2$ between adjacent zones. That is, a particular ring on the wavefront is $\lambda/2$ farther from the frontal point than the next inner zone and $\lambda/2$ closer than the next outer zone. Each of the rings has a small thickness, called a *Fresnel zone*.

One feature of the analysis is that the light from adjacent zones cancels each other out (for the most part) due to destructive interference. However, this cancellation isn't perfect because the electric field amplitudes are smaller as you go out to larger Fresnel zones. This feature helps explain Fresnel diffraction patterns in apertures. Figure 12-7 illustrates the arrangement for thinking about Fresnel zones.

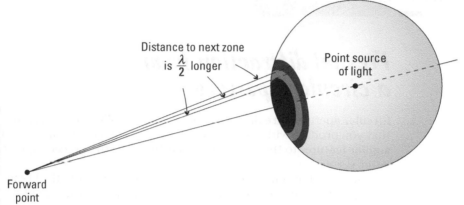

Distance to next zone is $\frac{\lambda}{2}$ longer

Point source of light

Figure 12-7: Geometry used to look at Fresnel zones.

Forward point

The following sections cover some special cases of Fresnel diffraction from several arrangements that occur in many imaging and other optical systems: a single slit, a circular aperture, a solid disk, and Fresnel zone plates.

Fresnel diffraction from a rectangular aperture

If you place a screen near a rectangular aperture in an opaque plate, a geometric shadow appears on the screen; however, you can see a complicated fringe pattern in the shape of the aperture inside the lighter area. Check out Chapter 11 for more on fringes. If you replace the rectangular aperture with a square aperture (in other words, reduce the length of the long side of the rectangular aperture), you get the same effect. An example for Fresnel diffraction for a square aperture is shown in Figure 12-8.

Figure 12-8:
Fresnel
diffraction
from a
square
aperture.

Fresnel diffraction from a circular aperture

Circular apertures are directly related to the details that you can see in images that you produce with most optical systems. Circular apertures produce similar features to the rectangular aperture (see the preceding section).

The source of the circular aperture's fringes comes from the Fresnel zones. The specific type of pattern that you see depends on the size of the aperture and the number of Fresnel zones that pass through the aperture. Because adjacent zones nearly (but not completely) cancel each other out, a small aperture creates a simpler pattern, meaning one or two fringes are visible. As the size of the aperture gets larger, more Fresnel zones pass through, and the pattern becomes more complicated, as you can see in Figure 12-9.

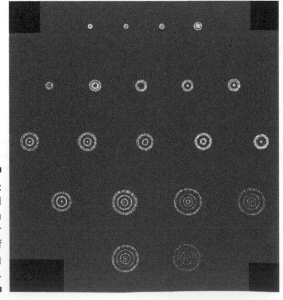

Figure 12-9:
Fresnel
diffraction
for circular
apertures of
increasing
size.

Fresnel diffraction from a solid disk

Augustine Fresnel entered his diffraction theory in a contest in 1818 in which he proposed a wave theory for light. A member of the judging committee, Simeon Poisson, pointed out that if Fresnel's theory was correct, a bright circular dot would appear in the middle of the *umbra,* or darkest part of the shadow produced by a solid disk. In this arrangement, the disk blocks the lowest-order Fresnel zones (the ones closest to the center) and allows only the higher-order zones (the ones farthest from the center) to pass. The rather interesting result here is that the higher-order zones create a bright spot in the center of the shadow, even though the brightest part of the light sent toward the disk is completely blocked. This spot is now called the *Poisson spot.*

You can get a general sense of this phenomenon if you have a C-cell or D-cell flashlight with a 2-inch-diameter or so lens. Turn on the flashlight, stand it on end (lens up), and place a quarter in the center of the lens. To see the diffraction pattern, hold a notecard close to the quarter. As you move the card away, you suddenly see a bright spot in the center of the quarter's umbra.

Diffraction from Fresnel zone plates

Diffraction and the Fresnel zones create another way to concentrate monochromatic light. If you block all of either the odd or the even Fresnel zones (basically, you block every other zone), the remaining light adds together via constructive interference. Luckily, Fresnel zone plates let you do just that.

Fresnel zone plates are special films or plates that block the odd or even order zones. The Fresnel zone plate in Figure 12-10 blocks the even orders.

Figure 12-10:
A Fresnel
zone plate.

Blocking every other Fresnel zone eliminates the cancellation effect caused by adjacent orders and results in a relatively bright light irradiance at the very center of the diffraction. If *collimated* light (light represented by parallel rays) is incident on a Fresnel zone plate, the light focuses at a particular distance, called the *primary focal point*. This result is similar to what a focusing lens does with light, but a lens uses refraction, not diffraction.

You can determine the primary focal length of Fresnel zone plates by the following equation:

$$f_1 = \frac{R_1^2}{\lambda}$$

In this equation,

- f_1 is the primary focal point of the Fresnel zone plate.
- R_1 is the radius of the first Fresnel zone in the zone plate.
- λ is the wavelength of the monochromatic light illuminating the zone plate.

You can see from the equation that the primary focal length depends on the wavelength of the light used. Unfortunately, this reliance means that Fresnel zone plates suffer from chromatic aberration (see Chapter 7) much more than lenses do. So although they can focus light like a lens, they work best with light of a single wavelength.

The radius of the first Fresnel zone is the radius of the relatively large circle at the very center of the zone plate. All other radii are measured from the center of the first Fresnel zone.

When you design a Fresnel zone plate, you first need to determine the focal length you want. With that value, you can calculate the necessary radius of the first Fresnel zone and then use the radius to calculate the radius of the other higher-order Fresnel zones, using the equation

$$R_m{}^2 = m f_1 \lambda$$

In this equation,

- R_m is the radius of the *m*th order Fresnel zone.
- *m* is the order number of the Fresnel zone.
- f_1 is the desired primary focal length of the zone plate.
- λ is the wavelength of the light used to illuminate the zone plate.

In addition to the primary focal length, Fresnel zone plates have other secondary focal points, but they have much lower intensities than the primary focal point. Because the light is concentrated by diffraction and not refraction, you can find bright spots if you start with a card at the primary focal point and translate inward. You find localized bright spots at distances of $f_1/3$, $f_1/5$, $f_1/7$, and so on.

Fresnel zone plates are particularly useful with light whose wavelength is too short for most readily available refracting materials to refract. For example, you can make zone plates with no material in the transparent zones. These plates can focus light in the ultraviolet to the soft x-ray region of the electromagnetic spectrum, where most refracting materials (glass and plastic) absorb this type of light.

Part IV
Optical Instrumentation: Putting Light to Practical Use

The 5th Wave
By Rich Tennant

MAGNI[FYI]NG
OP[TI]

"Oops."

In this part . . .

This part looks at how you can use the basic processes of manipulating light to make practical devices — tools that do something that you want. In this part, you see how eyeglasses correct for common vision problems and how other devices change the properties of images so that you can see objects that are far away or very tiny. The part also covers useful ways to produce light, including the development of a very unique light source, the laser. Finally, I discuss the basic ideas behind light guides and give you the scoop on using a fiber-optic link to send information.

Chapter 13

Lens Systems: Looking at Things the Way You Want to See Them

. . . .

In This Chapter

▶ Discovering some important optical properties of the human eye

▶ Understanding common vision problems and how lens systems correct them

▶ Reaching beyond the eye's natural capabilities with optical devices

. . . .

Sight is the primary physical sensor for human beings. Because sight involves light, optical systems have for many centuries played a very important part in improving the way humans can experience life. Although some basic devices (such as lenses and mirrors) and manipulation methods can affect what light does with images, they provide only a limited capability to do more than what the human eye can do alone. However, combining these elements in just the right way creates possibilities that far exceed the eye's limits.

In this chapter, I describe some of the optical characteristics of your primary optical system — your eyes. Because some people's eyes function differently in some situations, I present the most common basic problems with the human eye and explain how an eye doctor can help correct some of these problems. Finally, I discuss some optical arrangements called lens systems that can enhance sight.

Your Most Important Optical System: The Human Eye

Because most optical systems are built to aid human sight, looking at some of the basic features of the human eye makes sense. Human sight (the processing of the signals the eye sends to the brain) is a very complicated thing — much more complicated than just looking at how the eye forms an image — but I leave the neuroscience to the experts and discuss what I know about, which is the structure of the eye and how the eye adjusts the focus to form a clear image.

Understanding the structure of the human eye

Figure 13-1 shows the basic anatomy of the human eye, with the cornea at the forefront. The *cornea* is the primary light bender of the human eye. The cornea is thin, but its curvature and index difference from air are all that are necessary to cause most of the bending of the light rays. If you've ever opened your eyes while underwater, you know you can't see very clearly there even though you still have the same lens in your eye. The reason everything is blurry is that you've lost the second condition for focusing light; the index change from air to the cornea is larger than the index change from water to the cornea because water and the cornea have nearly the same index of refraction. Because the lens can't adjust enough to compensate for the loss of the light bending that needs to happen at the cornea, your underwater vision isn't very clear.

The cornea, not the lens, is the eye's primary imaging system. The lens is for fine-focus adjustment, not for forming an image on the retina.

The *iris* is the structure immediately behind the cornea and is the colored part of the eye with a black circle in the middle. The dark circle is the *pupil*, an opening to allow light into the lens and ultimately to the light detectors at the back of the eye. The primary function of the iris is to control the brightness of the image (or in technical terms, to limit the rate at which energy is delivered to the retina). If too much light hits your retina, your retina may be burned, giving you a blind spot. This risk is why looking at the midday sun is a bad idea; your iris can't close tightly enough to prevent your retina from getting burned.

The next structure is the lens. The *lens* is a somewhat-stiff structure made out of cells that look like the cells in the thin membrane in onion skins. The lens in Figure 13-1 looks rather like the biconvex lenses from Chapter 7. The overall lens structure is somewhat stiff but still slightly flexible and transparent to the visible spectrum of light. Connected to the lens are the ciliary muscles, which can change the focal length of the combined cornea and lens system by small amounts, so the lens (as modified by the cilliary muscles) is the fine-tuner for clear images (I discuss moving the focal point in the next section).

The thickness of the lens plays no role in focusing light rays. I talk about the two features that determine the focal length of a lens — the radius of curvature of the surface and the index change going from one medium to another — in Chapter 7.

Ideally, the cornea and lens cause the light rays to converge and form a clear image on the *retina,* the structure at the back of the eyeball that's in line with the iris and the lens. On the retina are two types of detectors, rods and cones. The *rods* are sensitive to light only and aren't capable of determining color. They're the light detectors you use when you're looking around at night or in low-light situations. The *cones* are able to determine color, but they require a large amount of light to generate a signal the brain can use. ***Remember:*** The cones are just the starting point for color perception; the way humans perceive color is very complicated and dependent on a lot of brain activity that's outside the scope of this book.

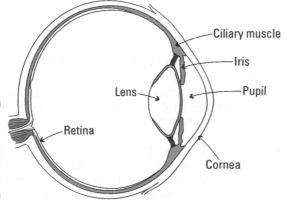

Figure 13-1:
The basic
structure
of the
human eye.

Accommodation: Flexing some muscles to change the focus

The process where the ciliary muscles change the curvature of the lens to produce a clear image on the retina is called *accommodation.* When the objects you're looking at are far away, such as a distant mountain range or clouds in the sky, the ciliary muscles are relaxed, and the lens is relatively flat, aside from its own slight, natural curve. For distant objects, the typical eye is relaxed, so it requires no accommodation to see them clearly. When you look at objects that are close, such as this page, the ciliary muscles are almost their most tense, and the lens has a decreased radius of curvature (decreased focal length). Looking at close objects typically requires near-maximum accommodation to produce a clear image on the retina. Objects that vary in distance between the two extremes (far away and up close) require different amounts of accommodation.

Chapter 7 shows the reason you need accommodation. If you approximate the cornea and lens as a thin lens system (it's not really a thin lens, but you get the gist of how it works if you approximate them combined as a thin lens), you can see from the thin lens equation that as you change the object distance, the image distance changes. Because the retina in your eye can't move, the focal length must change.

The degree to which an eye isn't able to accommodate either situation (objects far away or up close) is a symptom of some sight problems with the eye. Accommodation problems are physical problems, so optical systems can correct the physical situation. To find the limits of your eyes' capability to accommodate for objects at various distances, an eye doctor typically identifies two special points, the far point and the near point, for each of your eyes. After the eye doctor has determined that either of these points is out of the comfortable or useful range, she can prescribe corrective lenses for your eyes. I discuss vision problems and corrective lenses later in "Using Lens Systems to Correct Vision Problems."

The far point

The *far point* is the largest distance from your eye to the location where you can still see an object clearly. "Clearly" doesn't mean that you see all the details on the object, because the *aperture* (opening) of your eye is rather small (like two or three millimeters); rather, you can see the edges clearly. For an ideal eye, this distance is infinitely large.

In your eye doctor's office, infinity is hard to measure, so most eye doctors use a far point of at least 5 meters (about 15 feet), which is a reasonable distance for most everyday situations. So if you can clearly see an object that is 5 or more meters away from you, you have normal far vision. If you have to move the object to a distance of less than 5 meters, you may require glasses to see distant objects clearly.

The near point

The *near point* is the smallest distance from your eye to the location where you can still see an object clearly. The near point typically starts out relatively close (10 centimeters or so) when you're young; as you grow, this distance can move out to 20 centimeters at the teenage period and gradually increases the rest of your life.

Because most people like to do tasks, whether reading, building, or working with items, at comfortable arm's length, a functional near point is 25 centimeters (about 10 inches). If your near point is at most 25 centimeters, your vision is considered fine for daily activities. If your near point is greater than 30 centimeters, you may require glasses to get your near point back to 25 centimeters. Certainly, if your near point is, say, a meter away, you definitely want to correct that; you don't want to have to hold this book a meter away from your face to see the words clearly.

To check what your near point is, close one eye and hold a finger out at arm's length away from your open eye. Gradually move the finger toward your face until the edges begin to blur and you can't see your fingerprint grooves clearly. That's your near point.

Using Lens Systems to Correct Vision Problems

Several things can happen to the eye's anatomy that can affect how the entire system works. As you grow, the shape of your eyeball and cornea can change; an accident involving head trauma can change the shape of both as well. Using or not using the ciliary muscles can change their ability to accommodate, or adjust the focus for near and far objects. Most of the time, the factors that lead to less-than-ideally-functional sight are just natural defects in the structures.

Fortunately, some relatively simple optical devices can correct the eye's physical imaging problems. The following sections present three common eye problems and the corrective optical systems for them. But first, you need to know the way eye doctors talk about lenses and how those lenses affect light rays.

Corrective lenses: Looking at lens shape and optical power

If you look at a pair of eyeglasses or contact lenses, you notice that they don't look like the lenses presented in Chapter 7. I call corrective lenses *lens systems* because they contain two different types of refracting surfaces (and you usually need two lenses for your overall sight to work properly). Corrective lenses usually have a converging and a diverging surface fused together to form a *meniscus lens;* the diverging surface serves a couple roles like keeping the lens distance constant as you look around by rotating your eyeball and tends to reduce dis persion effects (where the index of refraction depends on wavelength, creating a different focal point for each color of light) created by the initial positive lens. One of the biggest problems with single refracting lenses is *chromatic aberration,* which means that dispersion causes different colors of light to focus at different spots. But this effect is unacceptable for corrective lenses because a partially focused rainbow image is as annoying as the original problem the lenses are correcting. You can see a cross section of a meniscus lens in Figure 13-2.

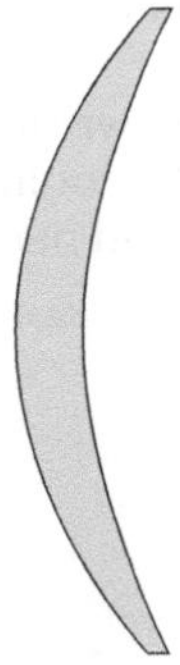

Figure 13-2:
A cross section of a meniscus lens.

Figure 13-2 demonstrates that the front of a corrective lens's surface bows out and the back surface bows in, away from the eye. So the meniscus lens is effectively one lens, and you can calculate its focal length by using the lens maker's equation in Chapter 7. The curved surface in back is what reduces the dispersion created by the first surface.

The reason for the curve that runs around your face (to the left and right) is that it makes the lens surface more or less equidistant from your eyeball, no matter what direction you're looking, without your having to turn your head. If you used a simple (single) lens, you'd only be able to see clearly through the exact center. Shifting your eyes without turning your head would make the image blurry, because the lens was designed for straight-on viewing.

Lenses are characterized by their *optical power,* their capability to bend the path of the light. The more a lens bends the path (the larger the refraction angle), the more powerful the lens. This bending power is known as the *dioptric power* of the surface. The units adopted for dioptric power are called *diopters*.

A diopter is equal to an inverse meter. If the focal length of a corrective lens is 10 centimeters, the dioptric power of the lens is +10 diopters. If the focal length of the lens is –15 centimeters, the dioptric power is –6.67 diopters.

Correcting nearsightedness, farsightedness, and astigmatism

Looking at the physical factors that affect image formation on the retina, three basic problems commonly occur that can be relatively easily corrected: nearsightedness (myopia), farsightedness (hyperopia), and astigmatism. This section describes these problems and how to correct them with lenses.

Nearsightedness (myopia)

As I note in the earlier section "The far point," you have trouble seeing distant objects clearly if you have a far point of less than 5 meters. Most likely, you can probably see close-up objects clearly, so you're regarded as being nearsighted. Doctors refer to this condition as *myopia.*

One early sign of nearsightedness is being unable to see the writing on the board in a classroom. If a child is sitting near the back of the room and feels the writing on the board is blurry, he has a short far point and needs glasses or contacts.

A variety of situations can cause you to be nearsighted:

✔ The curvature of the cornea or the relaxed curvature of the lens may be too strong, causing it to make light from distant objects focus in front of the retina and form a blurry image on the retina.

✔ The eyeball may be too long for the focusing power of the cornea and lens system.

✔ If you do a lot of work up close or like to read a lot, the ciliary muscles, like any other muscle in your body, become stronger and buffer. This strengthening causes the relaxed lens to have a little more curvature (making it produce a shorter focal length) than it would have without the strengthened muscles. Some people who read a lot, such as many college students, may become nearsighted because they're constantly using their ciliary muscles to focus on the words on the page.

To correct this situation, you need a lens that causes the image of a distant object to appear at your far point. With nearsightedness, the light rays focus in front of the retina, so the corrective lens needs to reduce the focusing power of the cornea and lens. A negative *(diverging)* lens does so quite nicely.

Correcting nearsightedness can improve other aspects of life, such as driving a car. Being able to see distant situations clearly can be very important to helping you drive without having an accident.

If you're diagnosed with nearsightedness, your eye doctor measures your far point and uses that distance as the focal length of the negative lens. The diverging lens still looks like a meniscus lens, but the curvature of the surfaces is adjusted so that both surfaces working together create a diverging lens.

Suppose your far point is 2 meters. Because this distance is too short for most of the situations you may be in while moving, especially driving, you need a corrective lens for nearsightedness. You can use the thin lens equation to determine the lens's focal length. In the calculations with the thin lens equation, objects that are far away are treated as infinitely far away — the object distance is infinite (or a very large number) such that the reciprocal of the object distance is zero. To get the distant object to appear at your far point, the object and the far point are in front of your corrective lens, meaning that you need a negative image distance. Setting up the thin lens equation, you get

$$\frac{1}{f} = \frac{1}{i} + \frac{1}{o}$$

$$\frac{1}{f} = \frac{1}{-2.0 \text{ meters}} + \frac{1}{\infty}$$

The focal length of the required corrective lens is –2.0 meters. Your prescription lists the dioptric power of the corrective lens as –1/2 diopters (the inverse of the focal length; see the earlier section "Corrective lenses: Looking at lens shape and optical power" for more on dioptric power). Using a negative lens to correct nearsightedness also pushes the near point back slightly. Nearsighted folks who need to do fine detail work, such as threading a needle, often find removing their eyeglasses useful to increase the magnification of the image formed on the retina. However, taking off your glasses is a quick fix; it reduces the magnification of the image, meaning you don't see fine details as easily. Just don't use this fix while you're driving! You shouldn't be threading needles while you're driving, anyway.

Farsightedness (hyperopia)

If you have trouble seeing objects that are a comfortable arm's length away, your near point may be less than ideal. (See "The near point" earlier in the chapter for more.) When this situation is the case, you typically can see distant objects clearly and are therefore farsighted (or, in doctor-speak, *hyperopic*).

Farsightedness is a little more difficult to diagnose than nearsightedness because most people adjust to farsightedness by pushing an object back until they can see it. Only when the near point becomes something like a foot and a half away does the problem become noticeable enough for people to want it corrected.

As a general rule, eye doctors treat the ideal near point as 25 centimeters from your face. (Optical system engineers use the near point of 25 centimeters when designing optical devices, which I describe later in this chapter.) If your eye doctor finds your near point to be something significantly greater than 25 centimeters, you may need corrective glasses to clearly see up-close objects.

Lots of factors can cause farsightedness:

- ✔ The curvature of the cornea or the relaxed curvature of the lens may be too flat, making light from distant objects focus in back of the retina and forming a blurry image on the retina.

- ✔ The eyeball may be too short for the focusing power of the cornea and lens system.

- ✔ As you age, the ciliary muscles may become less effective in bending the lens for near-object accommodation because the lens tissue becomes much less flexible with time. This situation explains why some older people use reading glasses to read the newspaper but then take them off to see everything else that isn't within the near point created by the glasses.

To correct this situation, you use a lens that makes the image of a close object appear at your near point. With hyperopia, the light rays aren't bent enough to focus on the retina and end up focusing behind it, so the corrective lens needs to increase the focusing power of the cornea and lens. The go-to optic for this job is a positive *(converging)* lens.

Finding the parameters of this lens requires the thin lens equation I cover in Chapter 7. Your near point distance represents the equation's image distance. The conventional ideal near point (25 centimeters) becomes the object distance. With the object distance and the image distance in the thin lens equation, you can solve for the required focal length of the positive lens. Although the curvatures of the surfaces work together to create a converging lens, the converging lens still resembles a meniscus lens.

Suppose that you're farsighted with a near point that is 130 centimeters. That number is the image distance for your corrective lens calculation. Because the object and your near point are in front of your lens and you need the image of the nearby object to appear at your near point, the image distance must be a negative number. In the thin lens equation, a negative image distance means that the image appears on the same side of the lens as the object. As I note earlier, the object distance is 25 centimeters. Plug those numbers into the thin lens equation:

$$\frac{1}{f} = \frac{1}{i} + \frac{1}{o}$$

$$\frac{1}{f} = \frac{1}{-1.30 \text{ meters}} + \frac{1}{0.25 \text{ meters}}$$

Solving for the focal length, *f*, you get +0.31 meters (31 centimeters). ***Remember:*** The positive sign is important because that tells the lens maker that the lens needs to be a converging lens. A negative lens would make your sight much worse! Your prescription from your eye doctor lists the dioptric power needed for the corrective lens — the inverse of the focal length, or +3.23 diopters.

Astigmatism (different focal points in different directions)

Perhaps the most common type of eye problem is *astigmatism*, which is the condition where the focal point of the eye in one direction occurs at a different point from the focal point of the eye in another direction. This discrepancy makes a blurry image that's tough to correct. To make matters worse, astigmatism may not be vertical or horizontal but some angle in between.

Laser surgery: Changing the shape of the cornea

You may have heard about a type of eye surgery called LASIK. Because the cornea is the main focusing element of your eye, changing the curvature of the cornea has a dramatic effect on the focusing power of the eye. LASIK can correct all types of eye problems. The principle is that the surgeon makes a calculated incision in the cornea. This cut may be to remove material from strategic locations to flatten the cornea (correct for nearsightedness) or to cause the formation of scar tissue that increases the curvature of the cornea (correct for farsightedness). The procedure can also improve astigmatism, but the calculations are much more complicated.

A major complication is that every person is different. Scar tissue formation and other healing factors can either overcompensate or not compensate enough for the particular problem a person has. This surgery works wonderfully for some people and not so well for others. Right now, predicting the exact outcome of the surgery is still difficult, but the surgical process is still improving.

The cause of astigmatism is an unusual curvature of the cornea. This curve can be a factor of growth (meaning the curve can further change, especially if you're a teenager) or the result of an injury. Whatever the cause, astigmatism itself doesn't fall into the farsighted or nearsighted category; in fact, you can have astigmatism on top of one of these other conditions.

To try to find the orientation of astigmatism, an eye doctor basically uses what's called a *cylindrical lens*. This lens focuses light in one dimension and then does nothing with the light in the perpendicular direction. The actual device the doctor uses is more complicated, but the concept of the cylindrical lens gives you an idea about how the process works. The eye doctor rotates the cylindrical lens around and changes its power to find what arrangement allows you to see an image clearly. When the doctor places a large metal plate in front of your face, flips and rotates some glass plates, and asks you which one looks better, he's using this system to diagnose astigmatism. With your particular astigmatism determined, the doctor notes the orientation of the cylindrical lens and the necessary focal length on the prescription so that you can get proper corrective lenses.

Enhancing the Human Eye with Lens Systems

The human eye is a remarkable sensor. However, sometimes your eyes just can't do a job as well as you want them to, even if your vision doesn't require corrective lenses (see "Using Lens Systems to Correct Vision Problems" earlier in the chapter). Using the proper combination of lenses and mirrors, many optical devices allow you to see details on tiny or faraway objects that you can't view with your eyes alone. The following sections discuss some of these devices.

Magnifying glasses: Enlarging images with the simple magnifier

To get a magnified view of an object with your naked eye, you bring the object closer to your eye. For example, if you bring your finger closer to one of your eyes, you notice that you can see your fingerprint more clearly and that the image appears larger. This pattern continues until the lens is no longer able to accommodate for the closeness of your finger, and your finger appears blurry.

A way to enlarge the image further is to increase the focusing power of your eye. You need a positive lens with a short focal length (large dioptric power) to enlarge the image on your retina, much like the corrective lens for hyperopia does. This positive lens is usually referred to as a magnifying glass or a *simple magnifier*. To enlarge the image, you place the object within your near point. The positive lens forms an enlarged image back at your near point. Figure 13-3 illustrates how a simple magnifier works.

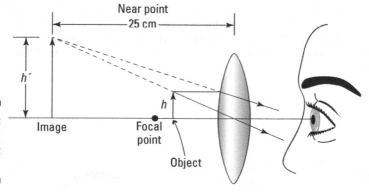

Figure 13-3: A simple magnifier at work.

A simple magnifier is characterized by its magnification power, usually denoted as a number with an ×, like 2×. The magnification power is a measure of how much bigger the image appears with the magnifier than it does at the retina with the unaided eye. The calculation for the magnification power is given by

$$MP = d_{np}D$$

In this equation,

- *MP* is the magnification power of the lens.

- d_{np} is the near point distance for the eye, typically taken as 0.25 meters. (Of course, that's the same as the 25 centimeters I use earlier in the chapter; I've left it in meters here because diopters are inverse meters.)

- *D* is the dioptric power of the lens.

For a normal eye with a near point of 0.25 meters and a typical magnifying glass with a focal length of 0.10 meters (dioptric power of 10 diopters) yields a magnification power of 2.5×.

A single lens magnifier is typically limited to a magnification power of about 3× due to aberrations. Magnifiers can have larger magnifications, but they're more complicated because they must use more lens elements to reduce the aberration effects.

Seeing small objects with the compound microscope

Because of greater aberration effects due to its small radius of curvature, a simple magnifier is limited in the amount of magnification power it can usefully produce, but a compound microscope can provide much greater magnification. The *compound microscope* uses two lenses placed in a certain way relative to each other to minimize the aberrations and increase the magnification power. You can see a schematic of a compound microscope in Figure 13-4.

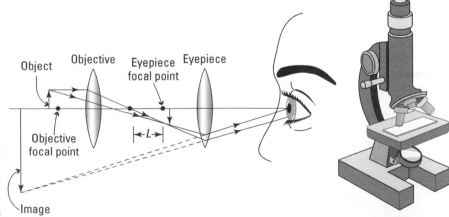

Figure 13-4:
The compound microscope.

The lens located closest to the object being examined is the *objective;* it's typically a very short focal length (large dioptric power) positive lens with a focal length less than 1 centimeter. The lens closest to the eye is the *eyepiece.* The eyepiece is, in the simplest case, a simple magnifier.

To use a compound microscope, you place an object slightly beyond the focal point of the objective to create a real inverted image in front of the eyepiece. The eyepiece is positioned such that this image forms near the focal point of the eyepiece. The eyepiece forms a virtual image that is enlarged even more than the first real image and inverted relative to the original object. **Note:** This inversion is why you have to think opposite when you adjust where the object lies under the objective when following moving bugs in pond water. If you need to move the image down, you have to adjust the microscope slide up.

This system's magnification is similar to the magnification-of-two-lenses concept in Chapter 7. The microscope's magnification is the product of the magnification produced by each lens. The magnification of the eyepiece (simple magnifier) is

$$MP = d_{np}D_e = M_e$$

I've changed the expression to M_e to indicate that it's the magnification power of the eyepiece. For most purposes, d_{np} is assumed to be 0.25 meters.

The magnification of the objective is given by

$$M_o = -D_o L$$

In this equation,

- ✔ M_o is the magnification power of the objective (the negative sign means that the image is inverted).

- ✔ D_o is the dioptric power of the objective.

- ✔ L is the distance between the focal point of the objective lens and the focal point of the eyepiece lens.

So, putting these two expressions together, the magnification of a compound microscope is

$$MP = M_o M_e = -D_o L(0.25 \text{ meters}) D_e$$

Aberrations still limit the magnification power of the compound microscope. More-sophisticated microscopes include expensive and complicated objectives and eyepieces that each contain multiple lens elements to correct for aberrations. However, light determines the real limit to how small an object you can see. You can't see features smaller than the wavelength of the light that you're using to look at the object. To see details smaller than the wavelength of light, you need to find a different wave. Enter *scanning electron microscopes,* which use electron waves that are much smaller than visible light waves.

Going the distance with the simple telescope

As you probably know, *telescopes* are optical devices designed to create a magnified image of distant objects so that you can see details more clearly. Telescopes come in two basic types: refracting (using lenses) and reflecting (using mirrors). I talk more about telescopes in Chapter 19. In this section, I examine the simple refracting telescope.

In a *refracting telescope,* two lenses are placed such that their separation is slightly larger than the sum of their focal lengths. Check out the simple refracting telescope in Figure 13-5.

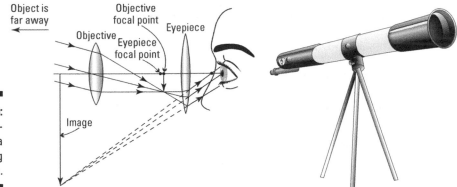

Figure 13-5: A schematic of a refracting telescope.

Like the compound microscope (see the preceding section), the refracting telescope has an objective and an eyepiece. Unlike the microscope setup, however, the telescope's objective is typically a long-length focal positive lens. Because the objects are far away, the light rays coming from the object into the telescope are basically parallel. The objective focuses these rays at its focal point, which is just within the focal point of the eyepiece. The image formed by the objective is real, reduced, and inverted. Because the image falls within the eyepiece's focal length, the eyepiece forms a virtual image that is enlarged and remains inverted relative to the original object.

The magnification of a telescope is given by

$$MP = -\frac{D_e}{D_o}$$

In this equation,

- ✔ *MP* is the magnification power of the telescope (the negative sign means that the image is inverted).
- ✔ D_e is the dioptric power of the eyepiece.
- ✔ D_o is the dioptric power of the objective.

In some applications, such as using a telescope to look at animals walking on a distant hillside, the magnification power of the telescope is important. In other applications, however, it doesn't make any difference. Stellar astronomy looks at distant objects, such as stars, that are so far away that reaching the level of magnification required to make their images look like anything other than dots is impossible. In this situation, scientists use telescopes with large objectives to capture as much light as possible; the light that travels over stellar distances is so weak that capturing enough light to even tell the objects are there is difficult.

Jumping to the big screen: The optical projector

An *optical projector,* such as the projectors you see in a movie theater or hooked up to a classroom computer, is like a telescope working in reverse. A small bright image is enlarged by two lenses and projected onto a screen.

The spacing of the lenses is like that in a telescope (see the preceding section), with the eyepiece (short focal length lens) placed closest to the film in a movie projector or the LCD screen in a computer projector. This setup generates a large image, but the depth of focus isn't very large. Therefore, the image on the screen appears blurry if the screen placement isn't exactly right or the screen isn't perfectly flat. Who wants to watch that? Using a second lens reduces the divergence of the image so that it's more controlled and has a larger depth of focus. In the telescope, the object is so far away that all the light rays from the object enter the telescope as parallel rays. The objective lens on the projector doesn't produce parallel rays. If it did, you wouldn't be able to see the image unless you had yet another lens to make it appear at the screen.

Another difference between a projector lens and a telescope is that the projector takes a small image and broadcasts a larger image on a screen instead of capturing a small amount of light from an object and concentrating it in a smaller image. Because a lens system doesn't add energy, enlarging an image weakens the image because the enlargement spreads the total energy over a large area. To make the image bright enough to see, an optical projector requires a light source. The larger the image made, the more light is required to make the image clear. For example, an IMAX projector requires a much brighter light source than the computer projector in a classroom.

Chapter 14

Exploring Light Sources: Getting Light Where You Want It

*O*n-demand lighting is probably one of the significant contributing factors to the development of civilization. Although torch and candlelight have been around for thousands of years, the light from these sources is rather dim, limiting what a person can do in the light they provide. Bright light sources such as the electric light bulb allow you to perform more-complicated tasks during the dark hours of the night (such as pulling an all-nighter before an exam) without the increased risk of burning your house down.

Lighting is used in many different ways, not just to light up a dark room. With the advent of environmental concerns, lighting has come under pressure from factors directly and indirectly associated with its particular application. No one lighting type can meet all needs, so the challenge is to find the most efficient method appropriate to its ultimate environment. Lighting engineers have a variety of materials and arrangements to choose from; recently, engineers have also benefitted from growing methods of sending light out from light sources, especially with display technologies such as flat-panel TVs.

In this chapter, I examine how common household light bulbs work, which covers how most lighting is done. In light (pun intended) of today's energy-conscious world, I also present some new lighting technologies that work more efficiently with the electricity given to them and discuss how optical engineers can use them in video displays. In addition, I discuss the most remarkable light source available: the laser.

Shedding Light on Common Household Bulbs

Most of the lights you use around your house are probably in bulb form. From a lighting engineering perspective, you can generally classify lighting for home and business into two categories: area lighting and task lighting.

- ✔ *Area lighting* is what most lights are used for: providing light in an area like a dark room, porch, or parking lot so that you can go anywhere in that area and generally see what you need to see.

- ✔ *Task lighting* is light provided in a concentrated area, usually where you do more-complicated tasks. Task lighting may be an overhead lamp for reading a book or preparing food in the kitchen for a meal.

Whatever the particular applications, the two most common types of bulbs current households use are the incandescent bulb and the fluorescent bulb. The following section presents the differences in the way these two bulbs operate. Later, I explain the units used to rate the amount of light electrical bulbs give off.

Popular bulb types and how they work

Light bulbs have been around for a long time. Incandescent bulbs were the first type of electric light bulb on the scene. Their relatively simple design and inexpensive manufacture allowed people of all walks of life access to this bright light source. As technology progressed, a more-efficient light bulb, the fluorescent bulb, came into being. This section presents the basic structures of these two types of light bulbs and the basic idea behind how they work.

Incandescent bulbs

Figure 14-1 shows a typical *incandescent light bulb*. This type of bulb produces light in a tiny wire called the *filament*. An electrical current (a stream of electrons) flows through the filament and produces atoms in an excited state by crashing into them. When the atoms de-excite, they can give off light; see Chapter 3 for more discussion about the light emitted by atomic transitions. Because the collisions produce a wide variety of transitions in the atom, they emit a rather broad range of wavelengths (colors) of light.

If you send a small amount of current through an unfrosted (clear, not white) bulb, which is basically what you do when you flip the light switch, you may notice that the filament emits light but the wires leading to the filament don't. This distinction happens because of an electrical property of materials called resistance, which measures how much opposition electrons encounter while they're moving through a material. The larger the resistance, the harder the time electrons have moving through the material. The difference between the filament and the wires leading up to the filament is the amount of resistance. The wires have a low resistance, so the electrons don't lose as much energy as they move through the wires. The filament has a relatively large resistance, so the electrons lose a lot of their energy as they collide with the atoms and electrons in the material. These energy losses heat the filament, causing it to glow (that is, give off light).

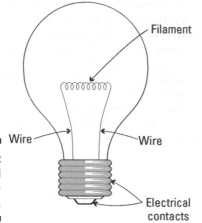

Figure 14-1:
A typical incandescent bulb.

Designers have to put some thought into the design of the filament besides just determining how bright the bulb should be. For example, the filament material is the main factor influencing the color of the light emitted, but designing the shape and size of the filament determines which colors, from those available in the material, are emitted. The geometry of the filament is what makes the light acceptable to you for your particular applications, such as soft light for your living room or grow lights for your plants. And of course, the filament must also last long enough to make it useful — filament life span was one of the main problems that Thomas Edison had in trying to get his incandescent bulb to work. Incandescent bulbs can work with either DC or AC electricity. This flexibility allows incandescent bulbs to work from batteries in flashlights as well as from your wall outlet for your desk lamp. This versatility also means you can use incandescent bulbs on a dimmer circuit.

Incandescent bulbs have gotten a bad rap lately, but they still provide a valuable function. The typical environmental problem associated with these bulbs is that they're considered inefficient because they actually give off more energy as heat than they do as visible light (which is what you most want from your light bulb). However, the heat incandescent bulbs generate can be useful. Incubators for baby animals such as chickens or turkeys use incandescent bulbs. Some people like incandescent bulbs in the bathroom because the bulbs can provide additional heat on chilly mornings. And in colder climates, city dwellers are learning that incandescent bulbs may still have a use in stoplights, where they melt the snow that builds up in the light housing. The new diode lighting in stoplights doesn't generate heat, so snow builds up and makes reading the light difficult.

Fluorescent bulbs

Fluorescent bulbs have also been around for a while, before the little twisty configurations you're probably used to appeared. These bulbs (see Figure 14-2) use an electrical current that collides with mercury gas in a sealed tube. The collisions cause the mercury atoms to go to a higher energy state such that when they de-excite, they produce ultraviolet light. This light is bad for your eyes, so the sealed tube is coated with a material to absorb the ultraviolet light and convert it to another color such as blue or yellow. The color of the light finally emitted by the bulb depends on the coating material. Usually, the application determines which material you use; lighting engineers go for materials that make soft white for general lighting, yellowish light for kitchen and bathroom lighting, and bluish light for plant grow lights.

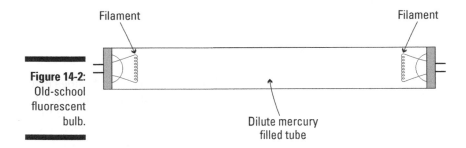

Filament   Filament

Figure 14-2:
Old-school
fluorescent
bulb.

Dilute mercury
filled tube

The fluorescent bulb is regarded as more efficient than the incandescent bulb because it emits very little heat, giving off more of its energy in the visible spectrum.

Fluorescent bulbs also require special conditions of the electricity they use. They need an electrical device called a *ballast* to condition the power. Because of this requirement, fluorescent bulbs work only with AC electricity and generally don't work well on circuits with dimmer switches.

Because fluorescent bulbs contain mercury, you have to be careful about handling burned out or broken bulbs. Make sure you dispose of them properly (don't throw them in your household trash) to prevent mercury contamination of the landfill.

Reading electrical bulb rates

When selecting a light bulb, you usually want to know how much light the bulb is capable of giving off. A couple of rating systems exist, giving you an idea about how much light the bulb emits, but usually you're only interested in the relative brightness — determining which bulb is brighter. The most common rating actually just gives you the electrical power consumed by the bulb, like many electrical appliances in your kitchen, but it doesn't give you the amount of visible light emitted. The other rating gives a better idea of the amount of light emitted, but it is unique to light sources, so it's not as common.

Incandescent bulbs are typically rated by their electrical power used. A 60-watt bulb (60 W) uses 60 watts of electrical power and dissipates 60 watts of power as heat and visible light. As I mention earlier in the chapter, much of this energy is heat, so the wattage isn't a good measure of how much light it provides. However, it does provide a good relative scale, in that a 100-watt bulb emits more light (is brighter) than a 25-watt bulb.

Fluorescent bulbs also list the power used, so you can easily compare them to incandescent bulbs. Until incandescent bulbs are phased out, you may notice on the packaging that the fluorescent bulb emits as much light as a 40- or 60-watt incandescent bulb; this designation helps you determine what kind of fluorescent bulb to replace your incandescent bulb with.

You may see other units, such as lumens or foot-candles, that actually relate to the amount of visible light emitted, especially as incandescent bulbs are phased out. This rating provides a much more accurate description of the amount of visible light emitted by a bulb, but usually only photographers, engineers, and electricians know about these units.

Shining More-Efficient Light on the Subject: Light Emitting Diodes

With the rise of concern about energy efficiency in lighting, new light sources that don't require crashing electrons into atoms are a hot commodity. *Light-emitting diodes* (LEDs) fit that bill: They involve electrons moving to a lower state and emitting light without collisions.

LEDs have been around for quite a while; the old red rectangular digit alarm clocks used this technology. LEDs are very efficient, converting nearly all the electrical power into visible light with very little heat. Although old-school LEDs primarily emitted light in the infrared and red parts of the spectrum, new fabrication techniques have allowed LEDs to emit light in the yellow and blue-green parts of the spectrum. The problem is that a single LED usually has too narrow a spectrum to be used as a general light source that you'd be comfortable with; creating reasonable home-lighting sources requires new ways of arranging different types of LEDs. The following sections give you a look behind the scenes of LEDs and their application in displays.

Because LEDs may contain arsenic, be cautious when dealing with burned-out or broken diodes. Don't handle them bare-handed or throw them in your household trash; that trash goes to the landfill, which can become contaminated.

Looking inside an LED

To make an LED, you first need a type of material called a *semiconductor.* You may be familiar with these materials in a computer context: Semiconductor devices called transistors are part of your computer processor.

Semiconductor materials used in electrical devices essentially come in two types: p-type and n-type. P-type semiconductors have a *dopant atom* (an atom that is different than the semiconductor material) that accepts an electron from a nearby semiconductor atom and creates an electron vacancy in the atom. Since this is a state that an electron would normally occupy but which has been taken by the dopant atom, this leaves an electron state vacancy called a *hole.* In p-type semiconductor materials, the location of the hole can be moved around, similar to an electron in an ordinary electrical circuit, so the material is treated as having a mobile positive charge. N-type semiconductors contain a dopant atom that gives up one of its loosely bound electrons to the semiconductor atom, thereby creating an extra electron that can move easily, more or less, in the material. In both cases, the semiconductor material with dopants remains electrically neutral, but the hole or electron can be moved around under the right conditions.

To make an LED, you need both types of materials, positioned so that one type is on one side, the other type is on the other side, and the two materials are touching. When you provide power to the circuit containing this semiconductor junction, the positive and negative charges flow and meet at the junction, combining to emit light. The energy difference between the positive and negative charges (dictated by the properties of the materials used) determines the energy (color) of the light emitted; more energy means bluer light, and less energy means redder that the light. These devices are called *diodes,* and they work by allowing the charges to flow easily in one direction and experience large resistance if you change the direction the charges move.

Think of diodes as one-way streets for charges. Diodes allow the charges to move easily in one direction (called the *forward biased mode*) but cause a large resistance if you try to make the charges move in the opposite direction (called the *reverse biased mode*). LEDs are usually designed to be operated in the forward biased mode, so pay close attention to the plus and minus signs when connecting electricity to your LED.

All aboard for material classification

Material classification is based on the electrical properties of the material. Namely, you're dealing with the resistance of the material, which tells you how much opposition a material presents to electrons moving through the material. The units for resistance are ohms (Ω). You have three basic types of materials based on this electrical property:

✔ **Conductor:** A *conductor* is a material that electrons can move throughout easily. Metals are usually good examples of conductors, which is also why they're used for wires to carry electricity (electrical current) from one point to another.

✔ **Insulator or dielectric:** In an *insulator,* electrons have a hard time moving through

the material. On an insulator, charges usually stay where you put them. Rubber is a good example of an insulator. Electrical engineers usually refer to insulators as *dielectrics.*

✔ **Semiconductor:** A *semiconductor* is a material that has electrical properties in between a conductor and an insulator. Semiconductor materials are sometimes used with a *dopant* (a different atom or molecule) placed in the material. The goal of adding the dopant is to create a material where you can control how easily charges move through the material. They can flip a switch, so to speak, to allow current to flow or not.

Changing the voltage applied to a semiconductor may change the energy values slightly, but not enough to change colors. Color change is most easily accomplished by finding materials that have a particular energy difference, usually referred to by engineers as the *band gap* of the materials.

LEDs are more efficient because they don't emit light by random collisions with atoms. The positive and negative charges meet in the junction and combine to emit light, effectively moving the electron down to a lower energy state with no collision needed. This process creates very little heat. You can design and build LEDs to emit light in a particular color so that nearly all of the electrical energy is converted to visible light.

One of the drawbacks with LED lighting is that the light LEDs emit is basically one color. You can combine several different types of LEDs to try to mix the right amount of colors, but getting the same color of light as the bulb types of lights is difficult.

Adding color with organic light emitting diodes

The *organic light-emitting* diode (OLED) is a relatively new type of LED. Regular LEDs made with typical semiconductor materials are crystalline — very rigid and brittle. They also emit light with a very narrow spectrum, which means a single color. The semiconductors in OLEDs contain carbon (hence the name "organic"). OLEDs operate the same way that regular LEDs do, except that the structure contains relatively long polymer chains that are doped with other molecules that add additional negative or positive charges. When you connect the OLED to a power source, the positive and negative charges meet at the junction and produce light.

Because the organic molecules have so many different energy levels for the electrons to move to, the light from OLEDs is more appealing because it has many more colors. You get the efficiency of LEDs with something closer to the light offered by light bulbs.

Besides increased color choices, OLEDs have other advantages over LEDs:

- **Easier fabrication:** After the desired polymer chains are identified, OLEDs are much quicker, less labor-intensive, and therefore potentially cheaper to make.

- **Flexibility:** OLEDs are flexible, so as the technology progresses, you may be able to go to a store and buy a roll of OLED lighting strips that you can cut to whatever size you want. Stick them on the wall or ceiling like wallpaper and just plug them in for the ultimate in custom lighting.

LEDs on display: Improving your picture with semiconductor laser diodes

One lighting application that's important to many people is display technology, and LEDs are something of a mixed bag in this setting. LEDs produce light efficiently, but they suffer from relatively low brightness and resolution. To make the high-resolution images you see on an HDTV, the light source must be very bright. The smaller the source, the brighter it needs to be in order for you to be able to see it. LEDs are *small point sources,* which means that the light from the diode expands quickly as you move away from the source. This characteristic has the tendency to reduce brightness. Another problem with LEDs is that emitting different primary colors requires different materials, and placing different semiconductor materials close together is a challenge. Making them brighter and close together limits the resolution of the LED display. This obstacle remains a challenge for high-resolution LED displays, although semiconductor fabrication technology continues to improve.

On the other hand, LEDs function quite well in the video display realm. LEDs have a rigid, periodic structure, so an engineer knows exactly what type of light is emitted from a particular location. This structure is a very important feature to make sure that the image appears the way you expect it to appear in your flat-panel display. Recent advances in OLED manufacturing have improved the films' lifespan as well as significantly reduced the irregularities that appear in the films.

Even with all the advantages of LED technology for lighting and displays, one other improvement increases efficiency and resolution for displays: semiconductor laser diodes. Many dynamic billboards use large arrays of laser diodes to make bright, easily read signs that can be quickly programmed and operated by computers.

Semiconductor laser diodes (or *diode lasers*) add additional structures to the diode junctions to send more of the light generated in the same direction and reduce the expansion of the beam so that the source appears brighter. The latter isn't as important for general lighting applications, but it is a big help with displays. I discuss the details of lasers in general in the following section. The main advantage of semiconductor laser diodes is that they're more efficient, meaning that they have a greater brightness than other light sources (sending more of the total emitted light in one general direction than other non-laser sources) but can operate with much less electricity.

Typical semiconductor diode laser fabrication techniques make diodes that emit their light out to the side. That is, the device is made by depositing layers of material in the vertical direction, but the light comes out the side in the horizontal direction. Therefore, using the diode lasers in a display requires cutting and reorienting them on another flat plate to get the light to travel toward you so that you can see it. One very promising technology to pay attention to is the *vertical cavity surface emitting laser* (VCSEL, pronounced *vik*-sell). These semiconductor laser diodes are built in the vertical direction like regular diode lasers, but the light comes out the top. This technique removes two significant steps from the display fabrication process, allowing fabricators to build the display array on its final surface, while providing all the advantages of traditional laser diodes.

One interesting potential application of VCSEL technology is the possibility of turning a whole wall or even a room into a TV. You'd simply go to the hardware store and buy video tiles (or possibly strips, if VCSELs can translate to a flexible film like OLEDs can) to build a display as big as you wanted. With the proper computer programming, you could make the walls disappear and replace them with a view of a beach or any other environment all the way around you.

Zeroing in on Lasers

Perhaps mankind's most remarkable engineered light source is the laser. *Lasers* are more than simply concentrated beams of light (those you can generate with a light source and a lens system). They're complicated systems of light projection. Although people don't typically use lasers for general lighting, these light sources work for almost everything else. Entertainment, manufacturing, measuring, imaging, evaluating, communicating: Each application uses the different characteristics of lasers to accomplish the specific tasks it's uniquely suited for.

In this section, I describe how most atomic lasers work and present some of the features that distinguish them from light bulbs as a unique light source.

Not all laser systems use atoms; some use *free electrons* (electrons that are not bound to an atom). This kind of laser is called a *free electron laser* (FEL), and it has some advantages over atom-based lasers, primarily in the areas of specific wavelengths and very high power.

Building a basic laser system

Albert Einstein came up with the basic idea of the laser back in 1915 while he was working on a problem in statistical mechanics. Most laser systems involve atomic systems, although you can use potentially any process to make a laser. Some processes are simpler than others, but basic laser systems involve light made by electronic transitions in atoms, which I cover in the following sections.

Looking at laser physics

In your physical science class, you learned about the Bohr cartoon for the atom. I call it a *cartoon* because the atom doesn't really look like what Bohr's model says, but it does give you an idea about how the atom works in terms of energy levels. When an electron moves to a higher energy state called an *excited state,* the electron must gain energy from either a collision with another atom or electron or absorb the energy from a photon of just the right energy to cause the transition to the excited state. The two processes important to the operation of a laser involve the electron moving from an excited state to a lower energy state: spontaneous emission and stimulated emission.

- ✔ **Spontaneous emission:** When an electron is moved to an excited state, the electron waits a certain amount of time called the *excited state lifetime* and then gives up its energy by releasing a photon equal to the energy difference between the excited state and the lower energy state. The photon is released in a random direction with no connection to the initial collision or photon that caused the electron to be in the excited state. Look at the right side of Figure 14-3 to see a simple picture of this process.

- ✔ **Stimulated emission:** When an electron is in an excited state and a photon of the proper energy (equal to the energy difference between the excited state and next lower energy state) comes along, Einstein proposed that the electric field of the photon causes the electron to wiggle (due to the electric field force). As the electron wiggles due to the oscillating electric field, it gives up its energy by emitting a photon (see Chapter 3 about the sources of light) and moves to the lower energy state. The incident photon shakes the energy out of the electron. The significance of this process is that the emitted photon is exactly like the first photon that initially hit the atom. Look at the left side of Figure 14-3 to get the idea behind this process and notice that there are two identical photons coming out, not one like with spontaneous emission.

So you can see the basic idea of the laser: One photon goes in, and you get two exactly identical photons coming out. This process is the concept of *gain* in a laser where you get more identical photons out than you put in.

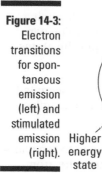
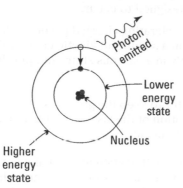
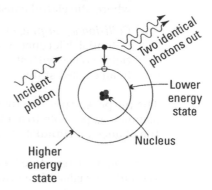

Figure 14-3: Electron transitions for spontaneous emission (left) and stimulated emission (right).

Getting enough photons to be useful requires many atoms. You can create a simplified three-level energy level system, shown in Figure 14-4, to show the processes involved in making a laser.

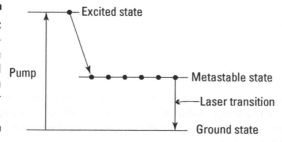

Figure 14-4: A simplified three energy level diagram of a laser process.

The energy levels pictured in Figure 14-4 are for a large collection of identical atoms. The bottom level isn't the lowest level in the atomic system but rather the highest energy level occupied by electrons in an unexcited atom. This level is usually referred to as the *ground state*. To create a situation where stimulated emission is possible, the atoms must be put into an excited state. This excitement is the function of the *pump* (represented by the large arrow on the left of the drawing in Figure 14-4), which is designed to take electrons from the ground state and put them in the highest energy state. The pump rate is large enough such that as soon as an electron appears in the ground state, the pump process puts it in the excited state.

The pump process comes in two basic types: optical pumping or collisional pumping.

- ✔ *Optical pumping* involves using light from a special light bulb called a *flash lamp* or from special laser diodes. This process uses relatively high-energy photons to produce the excited state. The higher energy pump light is used to avoid competition with the laser transition, where stimulated emission is designed to occur.

- ✔ *Collisional pumping* involves accelerating charged particles through a potential difference, giving them a significant kinetic energy. The particles collide with atoms, usually in a gaseous state, and put the atoms in an excited state.

The purpose of both methods is to try to keep the ground state unoccupied as much as possible to avoid absorption of the laser light and to maximize the chance for stimulated emission to occur.

The higher-energy state has a typical excited state lifetime of one to ten nanoseconds, meaning it's a practically instantaneous process where the atom casts off its excess energy and moves to the next-lower state. This next-lower state is special because its excited state lifetime is relatively quite long, perhaps on the order of microseconds to milliseconds. Although long, this state isn't permanent (because the ground state is the next lower state). This longer-lifetime energy state is important because you need atoms in an excited state in order to have stimulated emission.

This long-lived state is called a *metastable state*. A material usually has to have a metastable state for you to be able to use it to make a laser. Because the metastable state is relatively long lived, you can have more atoms in the excited state than in the lower ground state, known in a laser material as a *population inversion.*

Most systems in nature tend to configure themselves in the lowest energy state, which is usually the most stable. So the state where more atoms are in the excited state while a lower energy level is unpopulated isn't natural — it doesn't usually occur in nature.

Turning on a laser establishes a population inversion. Light spontaneously emits from atoms transitioning from the metastable state to the ground state. If this light is sent back into the material, stimulated emission can begin; the light is of just the right energy because it came from the atomic system being used. So one photon is sent into the material, and two identical photons come out (due to stimulated emission). If these two photons are sent back into the atomic system, four photons will emerge, and so on. This duplication makes all the photons in the laser exactly identical.

The number of photons in the laser continues to increase until nearly as many photons as atoms are present in the system. When this large wave of photons hits the population inversion, the entire population is de-excited by stimulated emission. This shift can overwhelm the pump so that significant numbers of atoms in the ground state begin to absorb some of the photons in the laser beam. The point at which absorption occurs is dependent on the pump rate, but regardless of the exact rate, all laser systems reach this point. This state where the power in the laser beam stays constant is called *saturation*. Saturation depends on many factors, but it explains why laser power doesn't continue to grow indefinitely.

Perusing the laser's parts

Most lasers (even those that aren't atomic) have four basic parts:

- **Gain:** The *gain* is the material where light amplification takes place. In an atom-based laser, the gain is where the population inversion is set up so the stimulated emission can happen.

- **Pump:** The pump is the mechanism that produces the excited state in the gain (see the preceding section for more on this process). In an atom-based laser, the pump can be a strong light source, like a flash lamp or laser diode, or an electrical current.

- **High reflector:** The *high reflector* is a 100-percent reflecting mirror that sends all the incident light that comes from the gain back into the gain to increase the number of identical photons. The high reflector is designed to reflect 100 percent of the light that comes from the de-excitation of electrons from the metastable state to the ground state. This mirror can be made with a metal coating, but using dielectric mirrors (made based on thin film interference — see Chapter 11 for more details) is much more effective.

- **Output coupler:** A laser is useless if you can't get light out of it to use. The *output coupler* is a mirror that allows a small amount of light out of the laser cavity and reflects the rest back into the gain. In most laser systems, this mirror is a concave mirror to increase the stability of the laser.

Figure 14-5 illustrates the basic schematic of a laser system.

Even though Einstein came up with the idea of the laser back in 1915, no one figured out how to make a stable system until 1959. The original idea was to use plane mirrors, but because light is a wave, it tends to diffract and spread out as it travels through openings or when reflected by circular mirrors; this diffraction, coupled with mechanical vibrations and changes caused by temperature changes, made plane mirrors really useless. Using a curved mirror, typically as the output coupler, made the laser possible because the presence of a focal point allows the mirror to compensate for both the natural tendency of light to spread out and for mechanical and thermal fluctuations by directing the incident rays back into the gain rather than out of the cavity like plane mirrors do.

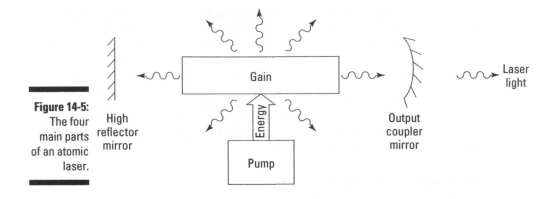

Figure 14-5:
The four main parts of an atomic laser.

Using a spherical element mirror in a laser cavity does have a consequence: You can't use the thin lens equation to predict what the beam will do when it hits a lens, because the beam isn't actually a ray. The beam changes size as it goes instead of traveling in a simple straight line. Rather, you have to use Gaussian beam matrix operations to model how a laser beam will be affected by a lens (and that's a topic for a laser text).

As you go from the middle to the edge of a cross section of a laser beam, the irradiance of the beam changes. The cross section of a laser beam is called the *beam spot* because if you were to look at a laser beam on a card, it would appear as a bright dot. If you look at the irradiance in the beam spot as a function of position in the spot, you find that it follows a bell curve distribution, or a *Gaussian distribution*. Because of this form of the irradiance of the spot, laser beams are often referred to as *Gaussian beams*. This Gaussian irradiance distribution has implications for what happens to the laser beam when you focus it. The beam doesn't go to a point but rather to a minimum size, called the *waist*. From the waist, the beam size changes nonlinearly; in fact the beam size as a function of position along the propagation direction follows a hyperbola. Figure 14-6 shows the beam size envelope as well as an example of the Gaussian irradiance distribution of a laser spot.

Figure 14-6:
The Gaussian distribution of a laser and the beam size as a function of position.

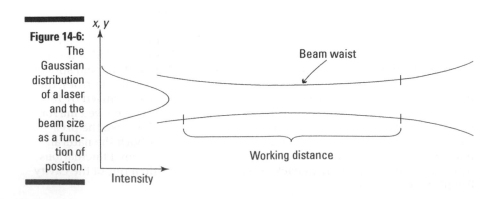

Notice in Figure 14-6 that the region right around the waist looks like it doesn't change size. This distance is called the *working distance* of the focused laser beam. It's important for some applications of lasers, such as drilling or cutting metal. I discuss applications of this feature in Chapter 18.

One last feature of lasers that you need to be aware of is the mode of operation. You can operate lasers in two different modes: continuous wave (CW) and pulsed mode. In *CW mode,* the laser is always on. This mode of operation produces a beam that looks the best and works best for imaging and measurement applications. *Pulsed mode* is where the beam is on for a period of time and then off and on. You use this mode when you need to get as much energy in the pulse as possible, such as in missile defense systems, or to precisely control the total amount of energy delivered to an area, such as in a surgical application where you're removing a cancerous tumor from healthy tissue.

Comparing lasers to light bulbs

The light produced by lasers is quite different than light produced by the other light sources I discuss in this chapter. Here are some main differences:

- ✔ **Directionality of the light:** Light from LEDs or light bulbs goes in all directions, which is why those sources work well for general lighting. Because of the laser's stimulated emission and mirrors, the light produced in a laser is concentrated in a rather narrow beam of light and travels in one direction. This directionality makes lasers rather useless for general lighting but allows them to work very well when you need to concentrate the light in a small area.

- ✔ **Brightness:** Because a laser's light is concentrated in one direction, the light is very bright. The brightness (irradiance) of lasers is something you can't reproduce with other light sources, even if you try to use lenses to try to concentrate the light.

  This brightness is why you should be very careful working with lasers (even laser pointers); the brightness can damage your retina very quickly.

- ✔ **Spectral purity:** Because the laser is made from atomic transitions and the mirrors that make up the cavity, the light in a laser is often referred to as *spectrally pure,* which means that relatively few wavelengths are present in laser light. Light from light bulbs has many different colors in it, like sunlight, which is important for general lighting that you're comfortable being in. Laser light is one color only. Spectral purity has advantages in some applications that other light sources would not work so well in, such as fiber-optic communication systems.

✔ **Coherence:** Before the laser, getting *coherent light* (light where the waves have a fixed phase relationship for a significant period of time) was difficult. Because laser light comes from stimulated emission (see "Looking at laser physics" earlier in the chapter), the photons are identical, so the phase doesn't change randomly or quickly. This consistency lets you use interference to make many measurements that can only come about with a bright, coherent light source. I talk about coherence in Chapter 11.

Chapter 15

Guiding Light From Here to Anywhere

● ●

In This Chapter

▶ Exploring how light guides use total internal reflection

▶ Looking at types of light guides

▶ Discovering light pipes and imaging fiber bundles

● ●

*L*ight guides are special structures that allow you to send light where you want it when you don't have a direct line of sight to the desired endpoint. These structures are responsible for revolutions in the data you can send, including video images, and in medical applications.

In this chapter, I describe the basic characteristics of some common light guides and present some typical applications of light guides. I also look at the differences between light pipes and imaging fiber bundles, and then examine how a fiber-optic communication link works and describe some performance characteristics that you may need to calculate.

Getting Light in the Guide and Keeping it There: Total Internal Reflection

One common characteristic for light guides is that they're relatively small. If you think of light guides as small wires, the cross-sectional diameters are typically smaller than 100 micrometers, or about the diameter of a hair on your head. Getting light into something this small isn't a trivial exercise.

Besides just getting the light into the small target, you need to make sure that the light stays put. All light guides work with the same principle: total internal reflection, which I describe in more detail in Chapter 4. To make a light guide work, you must make an arrangement where internal reflection can happen; namely, the light must start in a material with a larger index of refraction than the surrounding material (or second medium). As long as the light hits the boundary between the two materials at an angle of incidence greater than the critical angle, the light will remain in the high-index material and travel down its length, bouncing off the edges.

Navigating numerical aperture: How much light can you put in?

Figure 15-1 shows the basic arrangement for a light guide. Because the light guide works only if the rays strike the edge at an angle greater than the critical angle, you must send the light into the end of the guide at an angle smaller than a certain angle that I call θ_{max}. Figure 15-1 shows where a ray of light incident on the end at an angle θ_{max} refracts and hits the side of the guide with an incident angle equal to the critical angle, θ_c.

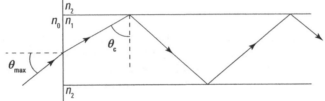

Figure 15-1: Light in the guide.

The following relatively simple equation allows you to calculate this maximum incident angle:

$$\theta_{max} = \sin^{-1}\left(\frac{1}{n_0} \sqrt{n_1^2 - n_2^2} \right)$$

In this equation,

- ✔ θ_{max} is the largest angle of incidence at which a ray can enter the guide and remain trapped inside the guide.

- ✔ n_0 is the index of refraction that the light guide is in. The material in question is usually air, which has a value of 1.00.

- ✔ n_1 is the index of refraction of the inner material, which is the larger index of refraction.

- ✔ n_2 is the index of refraction of the material that surrounds the inner material, which is the smaller index of refraction.

The θ_{max} expression also gives you another very important parameter for light guides: the numerical aperture (*NA*):

$$NA = n_0 \sin\theta_{max} = \sqrt{n_1^2 - n_2^2}$$

In this equation,

- ✔ *NA* is the numerical aperture of the light guide.
- ✔ n_0 is the index of refraction of the material the light guide is in.
- ✔ θ_{max} is the maximum angle at which a ray of light can enter the light guide and remain trapped inside the guide.
- ✔ n_1 is the index of refraction of the inner material, which is the larger index of refraction.
- ✔ n_2 is the index of refraction of the material that surrounds the inner material, which is the smaller index of refraction.

If you square the numerical aperture, you have a measure of the light-gathering power of the light guide. The larger the numerical aperture, the more light the light guide accepts. The numerical aperture can never be larger than 1 because the maximum angle cannot be larger than 90 degrees. If the right side of the equation gives a number larger than 1, the numerical aperture is set equal to 1. You can get commercial light guides (optical fibers, in particular) with numerical apertures of 0.2 to 1.0.

Examining light guide modes

As long as light is incident on the end of a light guide at an angle less than θ_{max}, it remains inside the guide. However, the particular angle at which the light enters the guide determines a particular path that is different from the other paths corresponding to different incident angles. Each path is called a *mode* of the light guide. The following list breaks down the basic classification of modes, which you can see in Figure 15-2:

- ✔ **Axial:** The path that goes right down the center of the light guide is called the *axial mode*. It doesn't bounce at all off the sides of the light guide, so it takes the least time to get through the guide.
- ✔ **Low-order:** A path that has a large angle of incidence on the sides of the light guide is a *low-order mode*.
- ✔ **High-order:** A path that has an angle of incidence near the critical angle is called a *high-order mode*. This mode bounces off the side of the fiber many times and so takes a longer time to travel through the light guide.

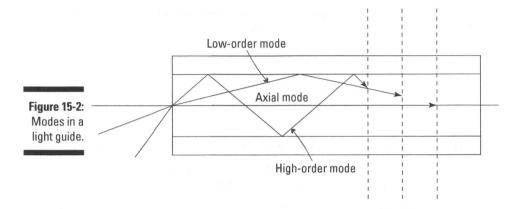

Figure 15-2:
Modes in a
light guide.

The consequences of the difference in how fast the light gets through the guide occurs when sending pulsed data in fiber-optic cables, which I talk about in the next section.

Categorizing Light Guide Types

You encounter two basic types of light guides: fiber optics and slab waveguides. Both types work with total internal reflection and function with light much like a pipe does for water. The pipe guides the water to where you want it to go. Light guides do the same thing for light. The following sections take a closer look at these divisions.

Fiber-optic cables

Fiber-optic cables are the most common type of light guide. The light is introduced into the fiber at one end and bounces around the fiber until it exits at the opposite end. Fiber diameters are typically less than 100 micrometers, so finding materials that are still rugged enough to work with this small of a diameter is difficult. The fibers are usually made of glass or plastic and look like very thin wires; glass is a common choice because it can be drawn into very long, thin strands and is somewhat flexible and heat-tolerant.

A fiber-optic cable consists of three main parts, shown in Figure 15-3:

- ✔ **Core:** The central part of the fiber that the light travels in
- ✔ **Cladding:** The layer of material around the core with a lower index of refraction than the core to ensure total internal reflection conditions are consistent throughout the length of the fiber
- ✔ **Buffer:** The colored layer of plastic around the cladding to protect the fiber from being scratched or bent too sharply

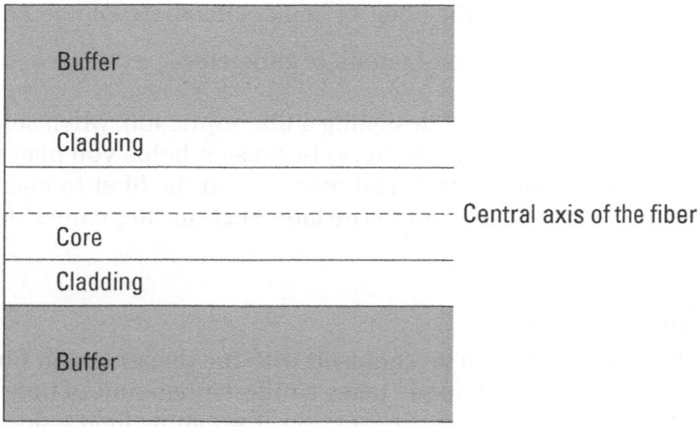

Buffer
Cladding
---------------------- Central axis of the fiber
Core
Cladding
Buffer

Figure 15-3: The parts of a basic fiber-optic cable.

Because fiber-optic cables are the long-haul light guide of choice for data communication, you have to consider several characteristics that significantly affect their use over long distances.

Attenuation

Because the core of a fiber is made of glass, impurities (such as iron, magnesium, or even water) and irregular structures can cause the light irradiance to decrease, a condition known as *attenuation*, as the light travels through kilometers of the core. The attenuation factor is well known for all the types of glass used in long-haul fiber-optic cables; you can find it in the fiber manufacturer's catalog.

The unit *decibel* describes the ratio of the optical power input into the fiber to the optical power measured at the output of the fiber of some length; it helps gauge attenuation. *Power,* which is the rate at which the light carries energy, is a more-convenient, more easily measured quantity than irradiance for characterizing the behavior of light in a fiber, so most fiber systems characterize the light-carrying capabilities of fibers by noting the effect on the power. The decibel is a log base 10 scale, so the number of decibels is equal to $-10\log(P_{out}/P_{in})$, which means that for a power ratio of 1/10, the measure is 10 decibels; 1/100 is 20 decibels, 1/1,000 is 30 decibels, and so on. The equation that tells you how much power you can get out of a fiber of a certain length is

$$\frac{P_{out}}{P_{in}} = 10^{-\alpha L/10}$$

In this equation,

✔ P_{out} is the power of the light exiting the fiber.

✔ P_{in} is the power input into the fiber.

    ✔ α is the attenuation of the fiber, in units of decibels/kilometer.

    ✔ L is the length of the fiber, in units of kilometers.

This equation is important for designing a fiber-optic link when sending data over long distances (tens of kilometers) because it helps you plan where you need to place signal amplifiers, called *repeaters,* in the fiber to make sure that the signal sent is still usable. Flip to the later section "Repeaters" for more on these devices.

Intermodal dispersion

Intermodal dispersion (not to be confused with the dispersion in Chapter 4) describes the fact that each mode takes a different amount of time to get to the end of the fiber. (See the earlier section "Examining light guide modes" for more on modes.) As I note in that section, the axial ray travels the shortest distance, so it takes the least amount of time to get to the end of the fiber. The higher-order modes have many more reflections off the sides of the fiber, so their rays have longer distances to travel and take the largest amounts of time to get to the end of the fiber. The time difference between the two rays traveling a distance of L in the fiber is given by the following:

$$\Delta t = \frac{Ln_1}{c}\left(\frac{n_1}{n_2}-1\right)$$

In this equation,

    ✔ Δt is the difference between the transit times for the highest supported mode in the fiber and the axial mode.

    ✔ L is the length of the fiber.

    ✔ n_1 is the index of refraction of the core of the fiber.

    ✔ n_2 is the index of refraction of the cladding of the fiber.

    ✔ c is the speed of light in vacuum, 3.0×10^8 meters/second.

This time difference, Δt, is important for data transmissions because it determines the minimum spacing required to keep the data pulses from blurring together. If you send a narrow pulse into the fiber, the intermodal dispersion tells you how much the pulse will broaden due to the time difference. If you place the data pulses too close together, intermodal dispersion blends them so that you can't distinguish the separate pulses, leading to a data error. To avoid this, you must separate the data pulses by a time larger than the intermodal dispersion. Although the actual time you use depends on many factors, the time difference due to intermodal dispersion is important to know because it's often the limiter for data transmission rates.

Material or spectral dispersion

Because the fibers are typically made from glass, the index of refraction depends on the wavelength of the light. For optical sources that emit many different wavelengths of light, each wavelength takes a different amount of time to travel through the fiber along the allowed modes. This time difference is called *material* or *spectral dispersion*. Although you can reduce this difference by using laser sources (which emit a narrower range of wavelengths), this effect is usually noticeable with minimal (but not necessarily nonexistent) intermodal dispersion, especially if the source is an LED. Another way to handle spectral dispersion is to find a range of wavelengths in the dispersion curve of the core material that doesn't change very much. Regardless of how you deal with it, the spectral dispersion is another factor that may limit data transmission rates.

Fiber type

Which fiber type you choose is based on both its cost and how much intermodal dispersion it can tolerate. Different applications can tolerate different minimum amounts of intermodal dispersion. You have three basic types of fiber-optic cables, listed here from least to most expensive and shown in Figure 15-4.

✔ **Multi-mode stepped-index fiber:** This fiber has a *stepped index profile,* which means that the index changes abruptly from one value to another as you go from the core to the cladding (see Figure 15-4a). The core diameter is relatively large, on the order of 50 micrometers or more, so it can capture a large amount of light, has many modes, and is relatively easy to fabricate. These fibers work best with short distances and slow data transmission rates.

✔ **Graded-index (GRIN) fiber:** This fiber has an index of refraction that is highest in the middle of the core and gradually decreases to the value of the cladding (see Figure 15-4b). This index makes the mode that travels the smallest distance move more slowly. The core sizes for GRIN fibers are typically between 20 and 90 micrometers, so the fibers have fewer modes than multi-mode fibers. These fibers work well over intermediate distances, such as between cities.

✔ **Single-mode fiber:** This fiber has a stepped index but a very tiny core, so only one mode can exist in the fiber (see Figure 15-4c). These fibers have the smallest core size, between 2 and 9 micrometers. Data rates can be quite high, and single-mode fibers are typically used for long distances, such as across states or countries.

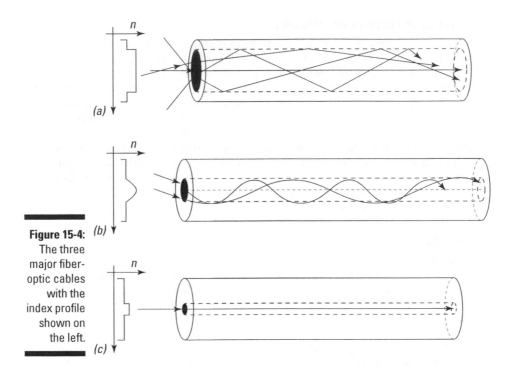

Multi-mode fibers are usually pumped with light-emitting diodes (LEDs; see Chapter 14), which are relatively inexpensive, but emit light with a large span of wavelengths. However, intermodal dispersion is significant in multi-mode fibers. You can reduce the intermodal dispersion a hundredfold by using a GRIN fiber. The modes in these fibers don't bounce off the edges; rather, they slowly spiral around the core. The modes on the outer edges of the core travel a larger distance, but the index of refraction is smaller, reducing the time difference between the axial and highest-order modes and therefore the intermodal dispersion.

The best solution to eliminate intermodal dispersion is the single-mode fiber because the very narrow core basically allows only one mode to exist. These fibers were typically stepped-index fibers, but newer ones today are GRIN fibers. Because single-mode fibers practically eliminate intermodal dispersion, the limiter of data rates in these fibers is spectral dispersion, even with laser sources.

Slab waveguides

Although fiber-optic cables are typically used to carry information over long distances, fibers don't work so well as *components* (relatively small devices that do things with light other than carry it from one place to another). Some components necessary for fiber telecommunication networks are things such as *modulators* (devices that put information onto the light) or *routers* (switches that send light into different fibers so that the information goes to the correct place). These devices use slab waveguides.

Slab waveguides have a rectangular cross section and are more easily made than cylindrical waveguides (which is what the glass fibers are) by using processes common to the semiconductor industry. Although fibers produce intermodal and spectral dispersion, these factors usually don't affect the small distances that components work with. Slab waveguide devices typically work with external voltages, so their performance is usually limited by the electronics connected to them rather than by any optical properties of the materials.

Slab waveguides can be useful for optical processing of signals, or carrying information around, say, a computer-processing or video-processing board. Because optical data transfer rates are typically much higher than electron-based data transfer rates, information-intensive operations may use slab waveguides rather than currents in wires to carry the information. When optical computers arrive, all information will be carried by such waveguides.

Putting Light Guides to Work: Common Applications

Light guides are very useful ways to get light where you want it to go, especially if the endpoint isn't viewable without them. Because of the flexibility of fibers, many applications have developed relatively recently that have dramatically affected technology and quality of life.

Light pipes

Light pipes are literally plastic or glass rods that can channel light. This arrangement has been known about for perhaps 100 years. The function of the light pipe is to emit light from a source exactly where you want it. Although light pipe technology has been used for art, it can also put light in dark places, such as a collapsed building or a body cavity, to provide a safer look at what's inside without risking a cave-in or infection.

Telecommunication links

The single largest application of light guides is *telecommunication links,* the fiber-optic cables that carry information such as telephone conversations or computer data. Fiber optics revolutionized the amount of data that people can transmit as well as the speed at which they can send it. The old copper wires used with electrical signals were much bulkier, and the phone conversations weren't really clear because of inductive effects. Telecommunication links can be small, like within a building, or large, like across a country. The increased data capacity of fibers came while significantly reducing the size of the cable used to transfer the information.

Figure 15-5 shows the basic components of a single link. Modern links are much more complicated, but this basic idea shown still applies. In the following sections, I describe the basic parts of the telecommunication link and then mention a couple of techniques that are commonly used to increase the data carrying capacity of modern fibers.

Modulator

A *modulator* is a device that converts an electrical signal to an optical signal. Putting a signal on light involves one of two methods:

- ✔ **Direct modulation:** *Direct modulation* uses the input electrical signal to directly modulate the light source, or turn the light on and off in proportion to the input signal. This method works okay for LEDs but is hard on diode lasers because it causes a lot of mechanical stress on the structures (as it heats up and cools down). (Chapter 14 gives you the lowdown on these light sources.)

  You typically use direct modulation with short-distance links and relatively slow data transmission rates. Direct modulation of a diode laser causes the laser's temperature to fluctuate as you turn the laser on and off. Diode lasers are subject to a lot of dynamic shifts in power, central wavelength, and range of wavelengths (called *bandwidth*) that they emit as their temperature changes. When you factor in the dispersion effects in fibers, these effects can result in an unstable signal over long distances.

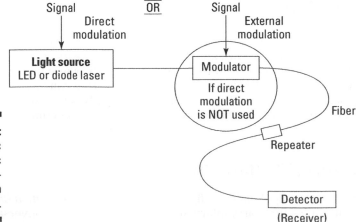

Figure 15-5:
A basic fiber-optic telecommunication link.

✔ **Electro-optic modulation:** The primary method to place a signal on a light beam is to use a separate modulator. A typical modulator in a telecommunication link is made in a slab waveguide (see the earlier section) with an electro-optic device in it. *Electro-optic modulators* change the index of refraction by applying an electric field. I describe these devices in more detail in Chapter 20.

Electro-optic modulators are used with lasers, usually diode lasers. The basic function of an electro-optic modulator is to change the power of the light sent into the fiber from the constant diode laser source, which is operated in its equilibrium state. The effect of the electro-optic modulator can be very quickly turned on and off, which produces a better signal than direct modulation of the laser, as well as higher data transmission rates.

Optical source and fiber

Two sources of light carry the data in telecommunication links: LEDs and diode lasers. Which source you use depends on many factors, such as distance, data transmission rates, and cost. LEDs are used for small networks and are the least-expensive source. Diode lasers are used for larger networks, including the intermediate- and long-distance networks. The diode lasers are more expensive, but they have the necessary characteristics to produce a clear signal after traveling through a long piece of fiber.

Which leads to the next piece of the link puzzle: the fiber. Head to the earlier section "Fiber type" for more on the factors that influence what fiber you use.

Repeaters

As I note earlier in the chapter, repeaters are signal conditioners and amplifiers that are necessary to overcome the signal attenuation that happens as light travels down a fiber. Using the attenuation equation and the dispersion calculations from the earlier "Getting Light in the Guide and Keeping it There: Total Internal Reflection" section, you can find exact positions where you need to install repeaters.

Repeaters aren't unique to fiber networks; the old electrical lines used amplifiers as well. With modern fibers, however, the spacing and, therefore, the total number of repeaters are much less that what the old electrical lines required.

Detector

Detectors are semiconductor devices, usually photodiodes, designed to absorb the light used in the fiber and convert the optical signal back into an electrical signal so that your telephone or computer can use the data. After optical computing is perfected, these devices will no longer be necessary. But for now, fiber-optic telecommunication links still need to produce an electrical signal at the output to work with current technology.

High data rates

Modern telecommunication links take advantage of the increased capacity of fiber-optic lines by using two techniques. Both of these methods facilitate the huge amounts of data transfer that you've become accustomed to while talking on your phone and surfing the Net.

The first technique is *time-division multiplexing* (TDM), which exploits the fact that communications such as phone conversations have rather large pauses in them. TDM fills those empty spaces with data from other signals so that the fiber is carrying as much data as possible at any given time.

The second method is *wavelength-division multiplexing* (WDM). WDM uses the fact that lights of different frequencies don't affect each other. Different signals are placed on different wavelengths, and a *multiplexer* combines them into one fiber. A *demultiplexer* on the other end separates out the different wavelengths, and a router sends them to their proper destination.

Imaging bundles

Light guides are useful for more than just bringing light to particular places or sending digital data. Fiber bundles can carry images from places that you can't see directly. A bundle is better than a light pipe (see the earlier section) for some applications because it's more flexible and can carry much more light than a single fiber.

Fiber bundles are made by placing several fibers close to each other in a bundle, and they come in two kinds: incoherent and coherent bundles. If the alignment of all the fibers is random, you have an *incoherent bundle,* which basically functions as a fiber light.

If the orientation of the fibers is maintained throughout the length of the bundle, you have a *coherent bundle.* This type of bundle can transfer images and is particularly useful when the object you want to look at isn't directly in your view. In the following sections, I explore two huge uses of coherent bundles: medicine and hazardous or secret situations.

Endoscopes: Medical imaging devices

Coherent bundles have made huge advances in medicine, especially in diagnostics and surgery. Doctors can now do laparoscopic surgery by making a few small incisions to permit a fiber light, an imaging bundle (called an *endoscope*), and surgical instruments into the body without opening the body and increasing the chance of infection. Chapter 17 shines more light on endoscopes.

Endoscopes have slight modifications for the particular types of body systems that they're designed to look into. The lengths of the endoscopes have control wires that allow them to move slightly, under direction of the doctor, so that the endoscope can be guided to look at particular areas of interest to the doctor. The endoscopes are also aided with a lens placed at the input end to generate a slight amount of magnification as well.

The fiber-optic snake: Rescue and reconnaissance

In nonmedical applications, a coherent bundle is called a *fiber-optic snake.* The bundles are often fitted with a lens at the input of the fiber, but the lens may be for increasing the field of view rather than providing magnification. A fiber-optic snake has control mechanisms along its length that allow the operator to move the snake in particular directions.

The fiber-optic snake is useful for looking into hazardous environments such as nuclear reactors, burning buildings, cave-ins, and so on. You may have also seen spies using snakes in movies to see what's going on in another room without having to be in the room. All this fun is brought to you by total internal reflection (see Chapter 4).

Part V

Hybrids: Exploring More Complicated Optical Systems

The 5th Wave

By Rich Tennant

"If you don't mind, my kid's got a new digital camera and he's looking for fun things to photograph for school..."

In this part . . .

This part looks at optical systems that have two or more basic phenomena incorporated into the device. These more complicated devices perform functions that affect many aspects of modern life. I show you the important parts of a camera and describe what they do; I also explain how you perceive depth and how holograms and 3-D movies reproduce this effect. You find out how optics is very much involved in medical imaging. In this part, I also discuss the many interesting applications provided by lasers. Finally, I present some different designs of telescopes, which have been used for centuries, that allow you to see the skies in different ways.

Chapter 16

Photography: Keeping an Image Forever

....

....

Knowing the properties of light and how to manipulate where light goes is one thing, but being able to build useful devices with this knowledge is another issue. One of the most common applications of optics, and one that you are probably quite familiar with, is photography: being able to produce and record an image.

Images are nice, but images made with lenses or mirrors change whenever the scene changes. Cameras, both still and video, save images somewhat permanently on some medium for you to enjoy for decades. Cameras, especially modern charged couple device (CCD) cameras, involve many principles of optics that must be coordinated properly to create the images you want to keep and to record them properly so that you can view them later. You get the best of both worlds: Cameras let you create the images you want to see and provide you with a mechanism for saving images that are special to you.

In this chapter, I describe the basic characteristics of both still and video cameras, including the old film style and the modern CCD style. I also explain how you perceive depth and how holograms and 3-D movies create this same effect.

Getting an Optical Snapshot of the Basic Camera

The basic camera arrangement is the same whether the camera produces moving pictures (videos or movies) or still pictures. All the parts are the same in terms of function, so the only difference is the rate and number of pictures taken. A still camera captures pictures one at a time; you have to press a button each time you want to take a picture. A video camera takes pictures at a relatively high rate of speed, like 30 per second, that results in the videos of birthday parties, family reunions, and viral shenanigans you're accustomed to watching.

Any camera has five basic parts (see Figure 16-1):

- ✔ **Lens or lens assembly:** Captures the light from a scene and forms an image at the plane of the image-recording medium.

- ✔ **Aperture:** Limits the rate at which energy hits the film to affect brightness.

- ✔ **Shutter:** Controls the time that light falls on the film, called the exposure time. In modern cameras, especially video cameras, this feature is replaced with automatic gain control (AGC).

- ✔ **Light-tight box:** Prevents light other than the image-bearing light that you want to record from hitting the picture-recording medium in the camera.

- ✔ **Recording medium:** Is a light-sensitive material or device that can store the image, such as photographic film or a CCD array.

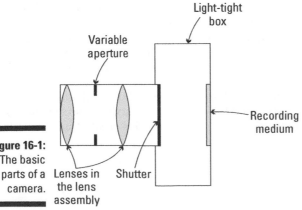

Figure 16-1: The basic parts of a camera.

Light-tight box

Variable aperture

Recording medium

Lenses in the lens assembly

Shutter

The following sections discuss most of these camera parts in more detail. Understanding how images are manipulated and stored with cameras provides experience with a very common application of forming images and controlling optical phenomena. This experience provides the basis for more-complicated image analysis (looking for details that are obscured by the effects of diffraction), called image processing (see Chapter 18 for more on this advanced application).

Lens: Determining what you see

The lens of a camera can enlarge or reduce the image on the recording medium and is the main part of the camera responsible for making a clear image (or not, when it isn't aligned properly). This section talks about some of the main types of lenses used with cameras, their characteristics, and the science behind focusing a lens.

Types of lenses: Telephoto and wide angle

The lens system that comes with your camera is an arrangement of several lenses designed to do a variety of things. The first thing the lenses need to do is compensate for chromatic aberration. *Chromatic aberration* occurs because the index of refraction depends on the wavelength of the light. (Head to Chapter 16 for more on chromatic aberration.) Lenses or lens systems that compensate for this effect are referred to as *achromats* or *achromatic lenses.* Most camera lenses are designed and built to compensate for chromatic aberration, which is one of the major contributors to the overall cost of a camera lens.

The next function of lenses is to change your field of view. The *field of view,* or *angle of view,* is how wide a scene you can see in the image. Sometimes, you need a wide field of view when you want to see a panoramic view of a mountain range; other times, you want a smaller field of view to concentrate on taking a picture of a family member, where you may not care for the background. These two situations create two extreme ranges for a lens to operate, so you actually have two lens systems to cover all the bases.

Telephoto lenses are lenses used to create a large image of an object or a person on the recording medium. The basic characteristics of telephoto lenses are

- Longer focal length
- Smaller field of view
- Greater magnification
- Smaller depth of field

This type of lens works well for portraits that allow for a comfortable distance between the camera and the subject. Cameras that don't have a telephoto range require that you physically place the camera very close to the subject; taking your portrait would be rather awkward if the photographer had to be only a couple of feet away from your face. If you have a small, pocket-sized camera, try taking a good portrait picture of a friend. Notice that you have to get very close if you want to be able to see his or her face clearly. You can also use telephoto lenses to enlarge images of portions of a faraway scene; you probably know this application as "zooming in."

Wide-angle lenses, on the other hand, capture more of a scene rather than concentrate on one particular subject. You use this type of lens to snap panoramic views like a mountain range or a beach scene. The images of individual objects are typically small, which allows many more of them to appear in the image.

The basic characteristics of wide-angle lenses are

- ✔ Shorter focal length
- ✔ Larger field of view
- ✔ Less magnification (reduction)
- ✔ Larger depth of field

Neither of these lenses would work in the variable day-to-day situations that you may want to take a picture of. Today's modern cameras have a zoom capability built in to their lens systems that allows you to be able to change between a wide-angle view (which provides relative magnifications from much smaller than one to values around one) to a view with a small amount of magnification without having to refocus the lenses. Zoom lenses have a set of lenses that can move inside the lens tube, which changes the effective focal length of the lens system to change the magnification produced. Zoom lenses are designed so that you can change the magnification of the image while keeping the image located at the recording medium. Because most people are interested in a compact camera, the relative magnification is typically limited to something like two to eight times. Larger cameras have changeable lens systems that accommodate much longer barrel lengths (lengths of the lens) so that a telephoto lens can provide larger magnification, but these most certainly don't fit in your pocket.

Focusing an image

One of the first things you probably discovered about dealing with cameras is that they can sometimes create fuzzy or blurry images. Although several factors can create a fuzzy picture, such the subject running by your camera very quickly, one big factor is the position of the lens in the camera.

In Chapter 6, I talk about the thin lens equation, which tells you where the image of an object will appear for a particular object distance from the lens. A clear, sharp image happens only when the object is exactly at the particular object distance; if the object moves to any other distance from the lens, you see a blurry image. Fortunately, you don't have to be as precise with your picture taking; you typically have ranges of object distances (depth of focus) and image distances (depth of field) where the image is sufficiently clear.

The old fashioned way of focusing an image shows you these two aspects of imaging, which relate to the location of the lens and the recording medium (see Figure 16-2).

✔ **Depth of focus:** As shown in Figure 16-2a, the *depth of focus* is the range of film locations where the resulting image is satisfactorily clear. As long as the recording medium is within this distance, the picture turns out okay, but if the medium is outside this range, the picture is blurry. Depth of focus is a nice camera feature because it allows you to be able to follow a slowly moving object without having to constantly readjust the focus.

✔ **Depth of field:** The *depth of field* (see Figure 16-2b) is the range of object distances that result in a clear image at the recording medium. When you have objects that are varying distances from you, like when you're looking at a scene with a lot of depth, having a large depth of field allows you to see an image of all the objects clearly. If you're focusing on two people talking in a mall hallway and you can see people clearly moving in a store behind them, you have a large depth of field. If you can see the two people clearly and everything else is blurry, the depth of field is shallow. Depth of focus is a major artistic feature professional photographers and movie directors use.

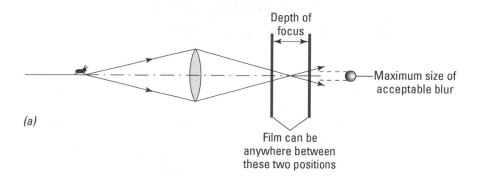

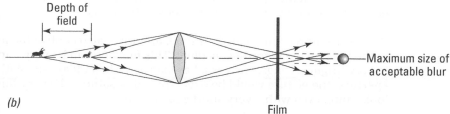

Figure 16-2: Depth of focus (a) and depth of field (b).

Real or processed zooming

If you read the specifications on your camera carefully, you may notice two types of magnification: optical zoom and digital zoom. *Optical zoom* is the magnification produced by physically changing the focal length of the lens system, producing a larger image of a smaller portion of the scene you're trying to photograph. On most portable cameras, the length of the barrel is very short, so you don't have a lot of room to move lenses around. Optical zoom is typically rather small.

Digital zoom uses computer programs to process the image to concentrate (or zoom in) on a particular part of an image. Because today's cameras have such a large pixel (picture element) count, computer programs can use this information to scale images relatively accurately to provide magnifications much larger than the short focal lengths required by the small lens barrel lengths of pocket-sized cameras allow.

To correct for blurry images, you usually have to make sure the camera is pointed at the object long enough for it to focus on the object (that is, create a clear image). Most cameras take care of focusing for you. In the old days, a photographer had to either adjust the location of the lens or the location of the film. Today's cameras focus an image by adjusting the position of the lenses so that a clear image forms at the recording medium.

Aperture: Working with f-number and lens speed

In the context of cameras, *apertures* are openings in opaque (nontransparent) sheets through which light passes, such as thin sheets of metal or plastic with a hole in them. The apertures in cameras are often called *stops* because they stop some of the incoming rays from entering the lens (light rays near the center pass through, while those on the outer edge are blocked by the opaque material). The adjustable apertures used in cameras are usually diaphragms made of several thin metal blades that you can move to change the size of the opening. As I mention in Chapter 6, apertures can reduce or eliminate spherical aberration, although at the cost of the amount of light that passes through the aperture. The smaller the aperture, the dimmer the image because fewer rays get through the lens to the image.

Besides reducing spherical aberration, apertures also control the irradiance of the light on the recording medium. This function can be very important in very bright settings, such as a baseball game in midday sun. Without the aperture, the picture would be overexposed or saturated; everything would look white, even with a very short exposure time.

f-number is the quantity that indicates the irradiance of the light incident on the recording medium relative to the irradiance entering the lens. It's not only related to the aperture size but also the focal length of the lens; it gives you an idea of the light-gathering power of the lens-aperture system. (Many photographers are probably more familiar with the term *f-stop*, which is defined the same way as f-number.) The f-number is defined as

$$f\text{-}number = \frac{f}{D}$$

In this equation,

- ✔ *f* is the focal length of the camera lens.
- ✔ *D* is the diameter of the aperture in the lens assembly.

The larger the diameter of the aperture, the smaller the f-number, which appears as *f*/f-number. If the focal length of the lens is 5 centimeters and the diameter is 1.25 centimeters, the f-number is 4. This notation appears on the barrel of the camera lens as *f*/4. If the aperture is reduced to 0.45 centimeters, the f-number appears as *f*/11.

A typical camera lens has a range of f-number arrangements available to you. The smallest f-number is referred to as the *lens speed*. The faster the lens, the more light the lens system can gather (and, typically, the greater the price of the lens because it requires extra work to compensate for chromatic and spherical aberrations).

The irradiance, *I,* of the light passing through the lens assembly is related to the f-number by

$$I \propto \frac{1}{\left(f\text{-}number\right)^2}$$

If you increase the f-number by some factor, the irradiance is reduced by the square of that factor. If you double your f-number, you get a quarter of the light passing though the smaller f-number arrangement.

On most modern cameras, software determines the proper f-number, so you don't really have to do anything with the aperture. However, for professional photographers, f-number can affect the depth of field. For the proper exposure time, the depth of field can be larger with larger f-numbers. Because the irradiance of the light is therefore reduced, the exposure time must be increased. As the aperture size is increased, the depth of field typically decreases.

Shutter: Letting just enough light through

A *mechanical shutter* is a piece of metal or plastic that prevents light from hitting the film until you're ready to take a picture. (Don't confuse it with the lens cover, which protects the lens from scratching in transport.) For film-based cameras, a timer in the camera times the shutter to allow just the right amount of light onto the film so that it's not overexposed. Pressing the button to take a picture activates the shutter; the shutter is usually responsible for the sound of a picture being taken.

In modern CCD cameras, a mechanical shutter isn't necessary. The information is taken from the CCD arrays, and the exposure effect (how bright the image appears) is controlled by *automatic gain control* (AGC). This software operation adjusts the amplification of the diode array so that light can constantly fall on the array without overexposure distorting the image unless you're dealing with a very bright object. AGC allows you to take pictures in a variety of light conditions without having to worry about the f-number or the sensitivity of the film. This technology has made it much easier to take pictures of just about anything without having to do any calculations or make adjustments to your camera based on experience. You can now just enjoy capturing that special moment, which is the point of having the camera anyway.

Recording media: Saving images forever

Cameras are meant to produce images that you can keep for long periods of time. In the old days, people used light-sensitive films — glass plates and, later, plastic, flexible films — that were coated with a silver halide emulsion that was sensitive to light (a change occurred in the emulsion based on the irradiance of the light on a particular area of the emulsion). The difference in the amount of light in the image formed on the film was recorded in the amount of blackness (think grayscale) created as the silver halide was converted to metallic silver, which appeared black on the film. The more light present, the more metallic silver was formed and the blacker that area looked; a dark area had little or no metallic silver, so it appeared lighter on the film. This process formed a negative image, one where the light and dark areas are reversed. The film was taken to a developer, who made the positive image that most people wanted.

Recently, CCD arrays (the image-recording medium in modern digital cameras) have replaced the old film-type cameras. CCDs are made up of small capacitors (electrical devices that can store charge) connected to a large number of very tiny electronic devices that produce an electrical charge in proportion to the amount of light that falls on them. The brighter the light that falls on a particular device, the larger the amount of charge placed in the capacitor. All the capacitors can send their information (the amount of charge they have stored) to a computer processor, which constructs an image based on how much charge was stored on the capacitor and where it was located in the array. Color images result from placing devices that have different color filters next to each other to detect the different colors present in the image sent to the CCD array. (Four devices with different color filters are grouped together to form a picture element or pixel.) Computer software renders the image based on this data.

The nice thing about the modern CCD cameras is that you don't need to take the film to a developer. You can transfer the images to a computer, do image processing on them, and print them out in any fashion you want. This convenience has basically driven film-based cameras out of business. Of course, if you don't have a computer or the hard drive crashes, you lose your pictures.

A feature of CCD cameras is the total number of pixels present in the array. Eight- to 15-megapixel cameras are currently available, and this density usually relates to how sharp of an image you can save. As this and earlier sections show, many factors contribute to the sharpness of an image; the large pixel number usually relates more to the ability to digitally zoom in on a picture and still have it appear clear and sharp than to just producing a clear picture.

Holography: Seeing Depth in a Flat Surface

Pictures are nice, but they look flat; they don't capture the appearance of depth that you're used to seeing as you look at objects in a room, which is usually most noticeable for objects that are relatively close to you rather than for faraway objects. *Holograms,* on the other hand, can not only show the variation in brightness of the image but also record the depth of the actual scene. The depth is usually called the *phase information* of the light that comes from the objects and helps you tell how far away the object is, especially when the objects are relatively close to you.

The concept of holography incorporates diffraction and provides a practical way to reproduce depth by recording the phase information of the light that travels different distances to the recording medium. Holography forms the basis of more-advanced image-processing techniques used with certain radar and medical-imaging technologies, as well as new diagnostic or quality control technologies used in product maintenance and manufacturing (some of which I cover in Chapter 18).

Seeing in three dimensions

Your two eyes are slightly offset, which means each eye's perspective is slightly different. Over time, you've learned how to use this difference to judge distance. This perspective difference is the feature that is missing from pictures. In regular two-dimensional pictures, each eye sees the same image;

you see only the difference in brightness. Therefore, the image appears flat. Holograms reproduce the effect from the actual objects at different distances from you. Each eye sees each object differently, and you perceive this difference as depth. The light that comes from the object toward your eyes differs slightly due to a different perspective angle.

To make a hologram, you need at least two beams. However, you don't want both beams to come off the object, because you don't have a large-scale reference to measure each angle. Without such a reference, each image produced washes any other image out. To solve this problem, holograms are made with at least one beam that reflects off an object and another beam that does not touch the object at all — it just touches the film. This first beam is called the *object beam,* and the second beam is called the *reference beam.* You can make a hologram with multiple object beams, but you can use only one reference beam. The reference beam provides the large-scale reference needed so that the image carried by each beam contributes to the overall image in a constructive way. In other words, the reference beam used during the film exposure process is what allows a clear image to be seen.

Exploring two types of holograms

You encounter two basic types of holograms: transmission and reflection; their main differences lie in how they're made and how you see them — that is, where the reference beam lies relative to the object and the film. These distinctions are important because you want to make sure you choose the right hologram for your situation. If you want to see a hologram in, say, the sunlight, you don't want a hologram that you can see only with the same laser light that made it.

Transmission holograms

You make *transmission holograms* with the object and reference beam on the same side of the film. Figure 16-3 shows a simple layout of optics to create a transmission hologram.

Notice that the reference beam touches the film only, avoiding the object. The light that bounces off the object and moves toward the film interferes with the light in the reference beam. Because the light in the reference beam travels in basically the same direction, the variation in the light from the object can be accurately recorded in the interference pattern that forms on the film.

After the film is developed, the light of the reference beam must be incident on the film for you to be able to view the hologram. You have to use the reference beam because the light passes through the emulsion. Because of dispersion, using light from a different light source doesn't make a clear image (because the different wavelengths refract with different angles), if one appears at all. When you use the original reference beam, the image is very clear and looks just like the object, with the depth based on the light that could travel from the object to the film.

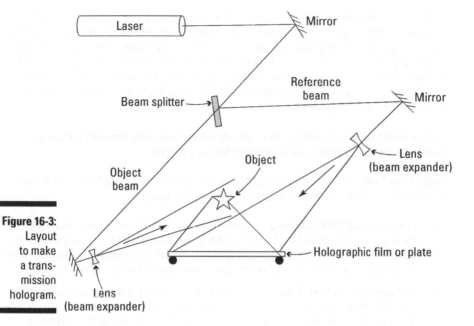

Figure 16-3:
Layout
to make
a trans-
mission
hologram.

Reflection holograms

Creating *reflection holograms* involves placing the object on one side of the film and the reference beam incident from the other side. You can see a simple setup for creating a reflection hologram in Figure 16-4.

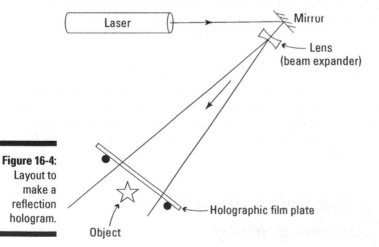

Figure 16-4:
Layout to
make a
reflection
hologram.

Notice that the reference beam is incident from the front, and the object is behind the film. In this arrangement, the light from the reference beam travels through the glass plate and onto the object. The light that bounces off the object and travels toward the film interferes with the reference light that is incident from the front. The interference pattern produced by the intersection of the two waves is so small that, in general, you can see the developed hologram in any bright light.

Because the light is traveling in opposite directions, the crossing angle is usually very large, resulting in a small fringe pattern.

Reflection holograms require slightly more-specialized considerations than transmission holograms do in order to work properly:

- **Transparent film backing:** Because the light from the reference beam must travel through the film, the film's backing must be transparent to light. In most settings, these holograms are formed on glass plates because of the extra stability and good transmission qualities the glass provides.

- **Smaller grain size:** Reflection holograms require a much smaller *grain size* (the average size of the individual silver halide crystals in the emulsion) than the transmission variety does; the interference pattern is much finer than the transmission case. This requirement makes the film for reflection holograms a little more expensive.

- **Protection from vibration:** Another consideration is that because the fringe spacing in reflection holograms is so small, it's less tolerant of vibrations. You have to take care to prevent vibrations from shaking the object relative to the light and the film so that the fringe pattern recorded on the film doesn't wash out. Most school settings use vibration-damped or vibration-isolation tables to minimize vibrations and keep from messing up the holograms.

Aside from the slight complications, reflection holograms are the ones students usually get to make in school, for the simple fact that you can see them in most bright lights, such as sunshine or the light of an overhead projector. However, if you have access to a helium-neon laser and a diffuser or negative lens to expand the laser light, you can see reflection holograms even more clearly as long as you condition your beam to follow the same path as the original reference beam.

Relating the hologram and the diffraction grating

A hologram is basically a very complicated diffraction grating. After development, the hologram diffracts the incident light along the paths that the light followed off the object to the film with the relative strengths that the light had after it bounced off the object. The end result is that the hologram changes the paths of the incident (viewing) light to follow the paths that the light made when it bounced off the object.

If you look at a hologram under a microscope, you see a bunch of tiny lines with seemingly random sizes and orientations. These lines are actually grooves formed in the developed holographic emulsion that are the result of an optical interference (fringe) pattern that was incident on the film. The spacing of the grooves works like a diffraction grating (flip to Chapter 12 for the lowdown on diffraction gratings). Any particular set of lines comes from the intersection of two coherent waves of light at some angle, relative to each other. The wave interference resulting from the intersecting light waves produces a fringe pattern consisting of a series of bright and dark bands (see Chapter 11 for more on interference and fringes). The spacing of the fringe pattern depends on the crossing angle between the light waves. Figure 16-5 gives you an idea about what the fringe spacing looks like with two different crossing angles between the pairs of waves.

Notice in Figure 16-5 that when the crossing angle is small, the spacing between fringes is large. When the crossing angle is large, the spacing between fringes is small. You get a constant variation between the two extremes such that the fringe spacing tells you the crossing angle between the two waves. This feature is the reason holograms can reproduce the depth associated with actual objects.

Because producing holograms requires an interference pattern, you must use coherent light. Light from a regular light bulb lacks coherence, so it doesn't work in this application. To make holograms, you usually need a laser (to provide an intense and coherent source of light) and a very mechanically stable table. That's why you don't have a pocket-sized hologram camera.

Figure 16-5: Fringe patterns produced by rays crossing at a small angle (left) and a large angle (right).

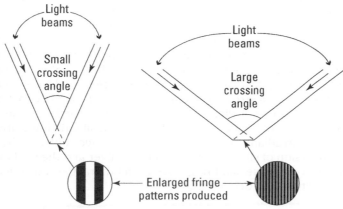

Enlarged fringe patterns produced

Making a holographic diffraction grating

You can form a very complicated diffraction grating when the fringe patterns are incident on a holographic film (as long as you take precautions not to overexpose the film). After the film is exposed, you develop it by using a process very similar to the process used to develop black-and-white film (which I discuss in the earlier section "Recording media: Saving images forever"). The main difference is that instead of preserving the metallic silver, you try to get rid of it in the hologram. So whereas a black-and-white film negative has dark and clear areas because the metallic silver is left in place, the hologram leaves the unexposed emulsion intact because the metallic silver is removed.

Part of the process fixes the unexposed silver halide so that it doesn't convert to metallic silver when you view the developed hologram in the light. After development, a slight groove forms in the emulsion where the metal silver was removed. The dark areas on the film have very little metallic silver, so this region is relatively unaffected during the developing process, making it look rather like a plateau.

Viewing a hologram: Depth from the diffraction grating

When light is sent onto the developed holographic film, the light interacts with the diffraction grating; light travels along identical paths to the original two light waves that made the pattern that was turned into the diffraction grating. Because the diffraction gratings in the hologram are generated by nearly all the light that is reflected off the object toward the film, the diffraction grating redirects the incident light along those paths. This process is how a hologram reproduces depth. Your eyes receive light from different parts of the hologram, which contains the image of different parts of the object. Just like when you're viewing an actual object, the hologram sends different information to each of your eyes, and you have the perception of depth.

The physical depth of the diffraction grating in a particular part of the hologram records the relative irradiances of the light rays that bounce off the object. The grating's physical depth depends on the irradiance of the light that is incident on the emulsion. The brightness or darkness of the fringe pattern depends on the relative irradiances of the two beams. If the two beams have equal irradiances, the fringe contrast is the best; that is, the bright fringes are their brightest, and the dark fringes are their darkest. If a weaker beam interferes with a stronger beam, the fringe contrast is very low; the bright fringes have only a slightly higher irradiance than the higher irradiance beam alone, and the dark fringes have only a slightly lower irradiance than the higher irradiance beam. The lower the fringe contrast, the lower the relative conversion of silver halide to metallic silver and the shallower the resulting grooves. These shallow grooves can't diffract as much light as a diffraction grating with deeper grooves.

Graduating to 3-D Movies: Depth that Moves!

Three-dimensional viewing of still pictures and movies has been around for quite a while. Because of the varied images and quick transition times, actual holograms aren't practical for movies or television screens. However, the basic idea — send slightly different images to each eye — can still apply. To accomplish this task, 3-D movies superimpose two images on the screen at the same time, or nearly the same time.

Human sight isn't a simple process, so just sending different information to each eye can cause problems, many of which we are just now beginning to understand. (The nearby sidebar "3-D is such a headache, and other problems" gives you more information on some of these problems.)

Movie and television screens utilize three processes for sending different information to each eye without using holograms: circular polarization, a six-color variation of the old two-color anaglyph system, and synchronized shutter glasses.

A significant drawback unique to these 3-D schemes that doesn't occur with holograms is that the viewer must be in just the right position to get the effect. If you like to do other things while watching TV, 3-D won't work for you because you won't be in the right position 100 percent of the time. 3-D works well in movie theaters because people sit in a particular position and, for the most part, stay there during the entire show.

Circular polarization

Some theaters still use the older polarization systems that employ linearly polarized light and linearly polarized glasses. Each lens allows linearly polarized light with one orientation to pass through and blocks out the light intended for the other eye. In principle, this setup allows one image into one eye and prevents the other image from being seen. (Chapter 8 gives you the lowdown on the types of polarized light.) However, experience in movie theaters shows that this method still causes *ghosting*, which occurs when the image meant for one eye leaks over into the image meant for the other eye. This effect is especially noticeable when the viewer tilts his head slightly. Circular polarization better minimizes the occurrence of ghosting, even with head tilting, but of course, the process is a little more complicated.

Circular polarization is generated in a digital projector as the image passes through the lens. It goes through a liquid crystal system that produces circularly polarized light. Circular polarization comes in two flavors, left and right. Passive glasses are designed to allow only one type to go through to each eye; left circularly polarized light only passes through one lens and right circularly polarized light through the other. This process creates the appearence of depth in the image.

The glasses for this type of 3-D viewing are relatively inexpensive, but the theater screen still needs to be a silver screen to preserve the polarization state of the light. Metals work because they reflect light without changing the polarization of the light incident on them. Linearly polarized 3-D theater screens also must be silver coated for the same reason, so a theater can upgrade to the better 3-D experience without springing for a new screen (but is still has to change the projector and the glasses).

A variation of this technology has been applied to certain large-screen digital televisions. A thin array of circular polarizers is placed such that half of the pixels are polarized for the right eye and the other half polarized for the left eye. This setup makes the TV a little more expensive, but viewers can use inexpensive passive glasses.

Six-color anaglyph system

Another advance with digital technology is the use of six-color anaglyph. This system is like the old red/blue images used in the 1950s to produce 3-D effects. In this modern version, light is separated into three colors and processed digitally to produce a true color image. Instead of using a whole color filter (again, like the old red/blue glasses), the anaglyph version uses two sets of three-color filters (a different set for each eye) that allow very narrow slices of red, blue, and green light through. Viewers use glasses with color filters that are matched to the projector to see the images.

The advantage of this system is that it doesn't require a silver screen. The special filter glasses cost much more than polarizing ones, though, so most theaters will want you to return them after the movie.

Shutter glasses

Shutter glasses are active glasses that have a liquid crystal display you can switch off and on in combination with a digital projector that alternately displays information for the left and right eyes separately. This alternation occurs so quickly that you don't notice the switch. The glasses are controlled by a radio frequency signal to synchronize the switching with the projector. However, these glasses are very expensive, and any slight problems with the synchronization can cause unpleasant results.

3-D is such a headache, and other problems

Human vision is complicated, and your sight is usually coordinated with other senses in your body. Sometimes, this coordination can lead to discomfort in 3-D applications. 3-D programmers use depth perception (sending slightly different information to each eye) to create the effect of something onscreen jumping out at you; they make the object appear close by creating a larger shift in the image that is sent to each eye. However, another process is in play as well: *vergence.* When objects are close to your face, each eyeball has to turn inward slightly to concentrate on the nearby object. The closer the object, the more inward they turn, resulting in extreme cases in something you may have done in elementary school: crossing your eyes. When you're focusing on a nearby object, your brain notices how much your eyeballs have rotated inward to see the object.

The problem with modern 3-D is that a mismatch occurs between the accommodation — the process your eye uses to form a clear image on your retina of objects close to you (see Chapter 13) — and the detected amount of vergence. This discrepancy can cause the nausea, headache, tired eyes, and other symptoms that sometimes happen with 3-D (much like motion sickness for some boat riders).

Another problem is that some people who view many hours of 3-D have problems seeing things in the real world for a while. This difficulty happens because your vision system adapts to the environment you're working in. The way to avoid this problem is to take breaks from the 3-D experience at regular intervals so that your vision doesn't adapt to your virtual world.

Chapter 17

Medical Imaging: Seeing What's Inside You (No Knives Necessary!)

● ●

In This Chapter

▶ Illuminating basic technologies for peering inside the body

▶ Checking out advanced ways to interpret the light your body emits

● ●

Most children have held a flashlight up to their hands and marveled at what they can see. Although that experiment is pretty rudimentary, the concept of using light (photons or waves) to explore the body has been around for quite a while. X-rays are a common example, but imagine being able to see other body elements, such as veins, arteries, cells, capillaries, and so on, all without making a single painful cut. Optics helps accomplish this feat.

As technology and the understanding of the atomic and subatomic world grow, science finds new uses of light to reveal what's hiding under your skin, and I cover some of these applications in this chapter.

Shining Light into You and Seeing What Comes Out

Although I use the word *light* in this book to talk about any electromagnetic radiation (because it all behaves the same way — it's just not all detectable by your eyes), most people in the medical-imaging field typically refer to light that is beyond the visible region of the electromagnetic spectrum as *EM radiation*. One of the more familiar and useful forms of this EM radiation is the x-ray. Just as light on the skin creates vitamin D or can give you a sunburn, other wavelengths, such as the x-ray, can have a predictable, measurable, and interesting effect on the body. I cover the x-ray and other useful methods in the following sections.

X-rays

X-radiation, more commonly known as *x-rays,* is a form of electromagnetic (EM) radiation shorter in wavelength than UV rays and longer than gamma rays. (Head to Chapter 3 for a look at the electromagnetic spectrum.)

X-rays allow you to see inside solid objects because the rays can penetrate solid matter and, depending on the density of the matter they encounter, pass completely through. Notice how different matter affects the amount of x-ray transmission in Figure 17-1.

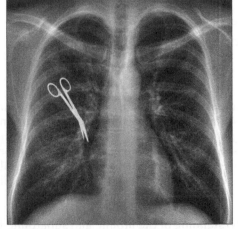

Figure 17-1:
An x-ray of the body and other common objects.

© istockphoto.com/Lukasz Panek

All EM radiation occurs naturally, but x-rays can also be man-made by using an x-ray tube. These high-voltage vacuum tubes accelerate cathode-released electrons to a high velocity; those high-velocity electrons collide with a metal anode, creating the x-rays. This method produces two kinds of x-rays, hard and soft, based on their wavelength:

✔ **Soft:** Longer wavelengths (10 to 0.1 nanometers) are *soft x-rays.* Soft x-rays don't have the ability to penetrate matter very far — less than 1 micrometer in water. Because soft x-rays don't contribute to producing the desired image, the machine filters them out to prevent increased radiation exposure of the subject. For the purposes of this discussion, I consider the application of hard x-rays only.

✔ **Hard:** The shorter wavelengths (0.10 to 0.01 nanometers) are hard *x-rays.* These are the rays I discuss in the rest of this section.

Wait, x-rays cause cancer too?

Medical science has identified a direct correlation between cancer and exposure to x-rays. X-rays do involve radiation, after all. The two important terms regarding the amount of x-rays an object is subjected to are exposure and absorbed dose.

✔ **Exposure:** *Exposure* is the measure of x-rays ionizing and is commonly expressed in *coulomb per kilogram* (C/kg). This figure is the amount of radiation required to create one coulomb of charge of each polarity in one kilogram of matter. You may also hear exposure discussed in the obsolete unit *roentgen* (R), which is the amount of radiation required to create one electrostatic unit of charge of each polarity in one cubic centimeter of dry air. One R equals 2.58×10^{-4} C/kg.

✔ **Absorbed dose:** The *absorbed dose* represents the effect of ionizing radiation on living tissue; the amount of energy deposited into the tissue is more important than the charge generated. The unit for absorbed dose is *gray* (Gy), the amount of radiation required to deposit one joule of energy in one kilogram of any kind of matter. Another, obsolete unit is the *rad*, which stands for 10 millijoules of energy deposited per kilogram. One hundred rad equals 1.00 Gy.

Each exposure to x-rays increases the risk of the patient developing cancer cells, so limiting x-ray exposure is a good idea. Scientists and engineers have begun to explore other optical technologies that provide the imagery of x-rays without the inherent risk.

Using x-rays to see inside a person requires an x-ray source and an image receptor (radiographic film), with the patient placed between the two. Short x-ray pulses illuminate the body or limb and either are absorbed or pass through, depending on the density and composition of the materials they encounter. As the x-rays leave the body, they strike the film. The film reacts to the amount of remaining x-ray that strikes it, with higher exposure producing dark areas on the film.

Bones show up as light areas on the developed film because they absorb more of the x-rays than soft tissue does. Blood vessels or the digestive tract are also visible if the patient has injected or ingested special x-ray blocking compounds, which absorb more of the x-rays than the surrounding tissue does. (Sounds fun, doesn't it?)

As technology has advanced, reusable, dynamic alternatives to the radiographic film have developed (no pun intended). These technologies (commonly called *digital radiology*) use photo-stimulated plates that react to the x-rays by storing some of that energy. Stimulating these plates with a laser causes them to release that energy as light. A computer can then measure and process this light to produce a digital image. This x-ray technology is what airport scanners now use.

Special optics can also collect, focus, and redirect x-rays, allowing you to make the rays go around a corner or obstacle. One such technology is called *polycapillary optics,* which are bundles of very small, hollow glass tubes. As x-rays strike the interior of these tubes at grazing incidence (similar to a rock skipping across water), the angle at which the ray strikes the tube walls is shallow enough that the ray doesn't penetrate the wall, so the whole x-ray stays inside the tube. The incidence angles at which the ray strikes the tube must be smaller than the critical angle, which controls the reflection of the x-rays down the length of the tube. The overlap of the beams from thousands of such tubes is what allows you to collect or focus the rays.

Optical coherence tomography

Optical coherence tomography (OCT) is an established medical-imaging technique widely used to obtain high-resolution images. It's well suited for applications, such as *ophthalmic* examinations (pertaining to the eye), that require micrometer resolution and millimeter penetration.

Unlike an x-ray (see the preceding section), OCT typically uses visible light to penetrate the body and provide an image with a resolution comparable to a low-power microscope.

This process requires several parts:

✔ The right kind of light source, such as super-bright LEDs, ultra-short pulsed lasers, or broadband white light.

✔ A means to split the beam. The Michelson interferometer, which I discuss in Chapter 11, is a common apparatus for splitting light beams.

✔ Two paths for the split beam to follow.

✔ A mirror to reflect one beam.

✔ A sample of interest (tissue) to reflect the other beam.

✔ A means to record and measure the interference pattern generated.

✔ A way to process the information from the pattern into a useful format.

A beam splitter divides the light into two paths, or *arms:*

✔ **Sample arm:** The *sample arm* interacts with the tissue. After it's inside the tissue (or other medium), most of the light scatters, and only a small portion of it reflects off subsurface features. (The scattered light has lost its original direction, creating glare that doesn't contribute to the desired image. The glare of the scattered light can cause the tissue to appear opaque or translucent.)

In the process of entering the sample and being reflected back through it, the beam of light changes. The wavelength of the light itself doesn't

> change, but the interaction of the sample arm with the tissue generates a phase change. The reflected light travels back to the beam splitter and is then redirected toward the photo detector.
>
> ✔ **Reference arm:** The other beam of light is called the *reference arm* and is used for comparison with the sample arm. This beam has the same basic properties as the beam traveling along the sample arm because its light was emitted at the same time. The beam of light travels along the reference arm, where it encounters a mirror that simply reflects the beam back to the beam splitter without changing the phase. The beam splitter then redirects the beam towards the photo detector.

As the beams recombine, the phase differences become evident and create either constructive or destructive interference (see Chapter 11). However, the only light that actually contributes to the interference pattern is the light from the two paths that have traveled distances within a *coherence length* of each other. The coherence length requirement means the photon packets of the beam need to have a well-defined (tight and consistent) length so that the only differences between the two beams are from the sample beam phase change.

The photo detector now sees the interference pattern generated by the combination of the out-of-phase beams and can record the interference. These interference patterns can be extremely small, so OCTs may use magnifying optics. Software can use the *fringes,* the widths of and spaces between the interfering lines, to extract information about the spatial dimensions and locations of structures below the surface of the sample.

The OCT process gives you a two-dimensional (depth into the tissue and width) image of a finite area of the sample. You can use scanning to turn this snapshot into a three-dimensional image by building up multiple pictures to create an image that includes length. Depending on the capabilities of the software, the scanner can move across the sample with images taken and recorded in real-time much like sonar or radar.

This method has limitations, however, and is only useful for imaging 1 to 2 millimeters below the surface of biological tissue. Too much of the light scatters (or is absorbed) at greater depths, leaving too little coherent light to be detected.

Endoscopes

Endoscopy typically refers to looking inside the body for medical reasons (though why you'd do it for another reason, I don't know). The process typically uses an endoscope to examine the interior of a hollow organ or cavity of the body. An *endoscope* is a medical device consisting of a long, thin tube with a light and an eyepiece or video camera. Unlike most other medical-imaging devices, endoscopes are inserted directly into the organ or body location of interest, with the light source outside the body and transmitted to the end of the scope. Doctors can see images of the inside of the patient's body through the eyepiece or on a screen. Lastly, the endoscope typically has an additional channel to allow for medical instruments or manipulators that enable biopsies and/or retrieval of foreign objects.

Endoscopes come in two basic types, rigid and flexible (see Figure 17-2):

✔ **Rigid:** The rigid scope uses a rigid tube of a specific diameter to transmit the light from outside the body by using special rod lenses. These rod lenses are optimized rigid glass rods that have special coatings and specific rod end curvature. Essentially, they're very long lenses. An objective transmits the lighted image from the end of the scope via the same type of glass rods to the eyepiece outside the body.

Rod lenses are generally used as light pipes (see Chapter 15) that can transmit light from one end to the other with very little light loss. These rods are very similar to the idea of fiber optics, except the rods aren't flexible. The rods are polished on the circumference and ground on both ends. Their optical performance is similar to a cylinder lens in that they focus collimated light passing through the diameter of the rod into a line.

✔ **Flexible:** The flexible scope replaces the rigid glass rods with a fiber-optics system. Controls in the operator's hands can steer the tip of a modern flexible endoscope. In the flexible scope, thin, flexible, transparent fibers act as the waveguides or light pipes. See Chapter 15 for more on fiber optics.

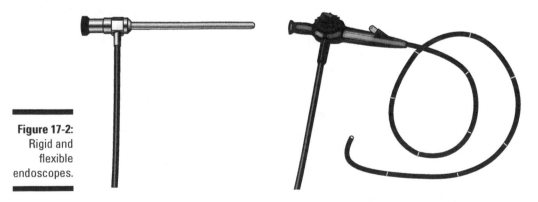

Figure 17-2: Rigid and flexible endoscopes.

Endoscopes come in a variety of configurations for a wide variety of applications. The nearby "So many -oscopies, so little time" sidebar provides a list of the more common endoscope types.

So many -oscopies, so little time

The endoscope is so useful that you can find one for just about any, well, "occasion." But seriously, because the human body is pretty sensitive to being pierced by such objects, endoscopes of different shapes and sizes allow doctors to minimize the trauma and discomfort of their intrusion into the body. Here are a few of the various uses for endoscopy:

✔ **Amnioscopy:** Examination of the amniotic cavity and fetus

✔ **Arthroscopy:** Examination of the joints

✔ **Bronchoscopy:** Examination of the air passages and the lungs

✔ **Colonoscopy:** Examination of the colon

✔ **Colposcopy:** Examination of the cervix and the tissues of the vagina and vulva

✔ **Cystoscopy:** Examination of the urinary bladder

✔ **EGD (esophageal gastroduodenoscopy) or panendoscopy:** Examination of the esophagus, stomach, and duodenum

✔ **ERCP (endoscopic retrograde cholangio-pancreatography):** Examination of the liver, gallbladder, bile ducts, and pancreas

✔ **Fetoscopy:** Examination of the fetus

✔ **Laparoscopy:** A small incision to examine the abdominal cavity

✔ **Laryngoscopy:** Examination of the back of the throat, including the voice box (larynx) and vocal cords

✔ **Proctoscopy:** Examination of the rectum and the end of the colon

✔ **Rhinoscopy:** Examination of the inside of the nose

✔ **Thoracoscopy:** Examination of the lungs or other structures in the chest cavity

Because endoscopes are basically a lens system, they don't require any type of image process unless they use a camera. However, the design of the fiber-optic system or rod lenses does require applying optical science.

Reading the Light that Comes Out of You

Remember that light is essentially electromagnetic energy (EM) that can act as either a particle or a wave. By exploiting and manipulating these properties, scientists have discovered ways to make materials and tissues send out information of interest. The science of optics (and not just the optical components but also the control and manipulation of light and other electromagnetic waves) in large part is making that possible.

This section explores the technologies that try to get the body to respond to probing by emitting light (or EM signal) to reveal the information of interest.

These signals can be processed and interpreted to find the location of certain tissues or chemicals and chemical processes within the body.

CAT scans

A *computerized axial tomography* (CAT or CT) scan operates very much like an x-ray (discussed earlier in this chapter). The primary difference is that the CAT procedure combines a number of x-ray images to create a more-complete picture. Using a computer, this machine can generate cross-sectional views as well as three-dimensional images of the internal organs and structures of the body, such as the folds of tissue in the brain shown in Figure 17-3.

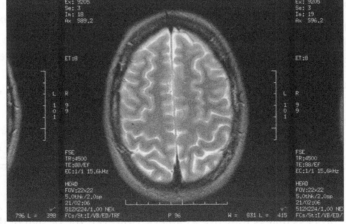

Figure 17-3: A CAT scan of a brain.

© Westend61/Getty Images

The CAT scanner looks like a large, donut-shaped machine. The hole in the middle accommodates the patient, and the "donut" houses the mechanism required to generate and record the images. In this machine, the x-ray emitter tube is on the opposite side of the hole from the x-ray detector, which is commonly based on photo diodes. As the x-ray photon strikes the photodiode, the detector records the photon's location and irradiance. *Contrast material* (an x-ray dye), often administered intravenously, can be placed into the spinal fluid to further enhance the scan and the various structural relationships of the spine, the spinal cord, and its nerves.

While the patient stands or lies still, a frame holding the emitter and detector rotates around the patient. The x-ray beam is focused to a thickness of between 1 and 10 millimeters and takes a full 360-degree rotation to make a single slice comprised of approximately 1,000 individual snapshots of the body. The detectors divide these snapshots into partitions, and then a computer backward reconstructs them into the two-dimensional image of that slice. After a slice of the subject has been scanned, the subject or machine can move axially to take a second slice. As the machine continues to move, the slices add together to form a complete and meaningful image of the body areas of interest.

Imagine the body as a loaf of bread. Each x-ray image slice is the equivalent to a single slice of the loaf. As you remove the end slice, you can see the inside of the loaf from crust to crust. Taking away slice after slice lets you see the whole inside of the loaf, including whole grains or the big pockets of air hiding in the middle. The CAT scan creates the same effect without having to actually slice through the body. (Donuts, bread; who knew CAT scans could be so appetizing?)

PET scans

Positron emission tomography (PET) is an imaging technique capable of producing a three-dimensional image of functional processes in the body. Based on nuclear imaging, this system detects pairs of gamma rays emitted indirectly by a positron-emitting radionuclide (tracer).

Taking the PET scan

To begin the process, the doctor introduces a tracer into the body on a *biologically active molecule* (something the body can process), such as FDG, an analogue of glucose. After a waiting period while the active molecule becomes concentrated in tissues of interest, the subject enters the imaging scanner. As the tracer decays, it gives information regarding metabolic activity and tissue concentrations within the body.

The tracer undergoes *positron emission decay,* emitting a *positron* (an antiparticle of the electron with opposite charge) that travels through neighboring tissue for a short distance, losing kinetic energy until it slows down enough to interact with an electron. This interaction destroys both electron and positron, producing a pair of photons moving in nearly opposite directions. The photons move away from each other but toward the *scintillator* (a material that emits light when excited by ionizing radiation) in the scanning device. Photomultiplier tubes detect the photons from the scintillator.

Successful PET imaging depends on simultaneous detection of the pair of photons. Because the photons are created simultaneously and moving in approximately opposite directions, you can localize their source along a straight line sometimes called the *line of response* or LOR. Because the point of photon emission isn't located exactly at the axis of the machine, the photons don't arrive at the scintillators at exactly the same time. To compensate for this discrepancy, the machine allows a few nanoseconds to register each pair of photons. If the time difference between registration of each photon in the pair is outside the allotted time window, the computer ignores them.

Reconstructing the PET scan

PET scan image reconstruction is very similar to the CAT scan reconstruction in the earlier section "CAT scan." The primary difference is that the PET process isn't as data-rich, so you have to use more-complicated processing techniques to obtain an acceptable image.

PET scans are also susceptible to (and must correct for) random events, such as scattered photons, detector dead-time and sensitivity changes. As such, PET scans require a considerable amount of data pre-processing. After the random events are filtered out, the remaining data is put into a set of simultaneous equations that represent the total activity of a parcel (section) of tissue. You can solve the equations by a number of techniques and then reconstruct the data based on how the photons were originally detected. Each ring of detectors can create an individual two-dimensional image, and then those stacked images can create a three-dimensional image. Alternately, all the rings can be used together to generate the three-dimensional image.

The solutions to the equations are used to create a map of radioactivity for a specific area of tissue. The resulting map shows concentrations of the molecular tracer in the tissues that a nuclear medicine physician or radiologist can interpret.

Although PET scan information is useful, it's only metabolic in nature and doesn't give the medical professional a complete picture of the body to coincide with the metabolic information. To resolve this issue, modern scanners often couple the PET scan with a simultaneous CAT scan in the same machine. Combining these images provides the complete metabolic image, overlaying the physical tissue images of the body.

NMR scans

Nuclear magnetic resonance (NMR) imaging is a premier method able to determine specific chemical structures found in fluid and tissue samples taken from the body. NMR is also commonly applied to the analysis of gases and other organic fluids.

NMR takes advantage of the fact that nuclei of atoms have magnetic properties that can be manipulated to yield chemical information.

Specifically, the *nuclear magnetic moment* (arising from the spin of neutrons and protons in an atomic nucleus) of an atom's nucleus will align with an externally applied static magnetic field. This applied magnetic field has a specific direction, but the rotating axis of the nucleus can't align exactly with the direction of that field. Instead, the rotational axis of the nucleus gyrates about some angle and a specific angular velocity. Not every nucleus will take on the exact same alignment, and these different alignments have different energy states.

A single transmitter generates a very short pulse of radio waves at a specific frequency (EM radiation in the MHz frequency range). This pulse actually creates a very wide band of frequencies. As the material sample is exposed to the radio waves, the protons in each nucleus absorb some of that energy and move to a higher but less favorable energy state (except when the atom is hydrogen, in which case the proton [nucleus of hydrogen] interacts directly with the radio waves). The absorption of this energy is called *resonance* because the frequency of the applied radiation and the gyration coincide or resonate.

But when the radio pulse ends, the nuclear magnetic moments are left in a nonequilibrium state of higher energy. In an effort to return to the equilibrium of the former lower energy state, they begin a process called *relaxation,* which generates an emission signal called a *free induction decay* (FID) that the instrument records. The nuclear magnetic moments are now back in equilibrium with the applied magnetic field.

In NMR, the energy loss that allows the nuclei to return to their low-energy state must go somewhere. One place is into the molecular framework (the *lattice*), where it's lost as vibrational or translational energy. The half-life for this process is known as *spin-lattice relaxation time* (T1). The other option is called *transverse relaxation energy*, where the energy is transferred to a neighboring nucleus. The half-life for this process is called the *spin-spin relaxation time* (T2). This process exchanges the spin of nucleus A with the spin of nucleus B, with no net change in spin. The values of T1 and T2 dictate the relaxation time of the scan, which can stretch into minutes.

The same coil that generated the radio signal now picks up the oscillating voltage generated by the net effect of the nuclear magnetic moments of all target nuclei in the sample relaxing back to equilibrium. This emission signal contains the sum of the frequencies from all the target nuclei; the signal is considered a high-frequency signal and is mixed with a lower-frequency signal to produce a digitized interferogram. A spectrum of the frequency domain (directly related to energy) is created by a Fourier transformation of the interferogram. These frequency domain plots are what help identify specific chemicals or compounds.

Doctors can obtain a wealth of information from an NMR spectrum. This technique has become a favorite among chemists because they can use it for identifying a product of a chemical reaction or determining the structure of a substance isolated from natural sources.

MRI scans

Magnetic resonance imaging (MRI) is a tomographic (*tomos* means "slice" in Greek) imaging technique capable of producing detailed images of internal physical structures and chemical characteristics. These images can be two-dimensional sectional images or three-dimensional volumetric images.

The change in image type or plane of view doesn't require mechanical adjustments to the imager.

MRI is a specialized (that is, more-complex) application of the NMR scanner (see the preceding section). NMR can be to analyze the structures of molecules such as proteins but is typically used on samples of relatively small size. MRI, on the other hand, is used on living subjects (human and other) and consequently must be able to physically accommodate the whole body. Although both NMR and MRI detect the presence of hydrogen in the sample or subject, MRI focuses on hydrogen in water. MRI is known for providing good contrast between the different soft tissues of the body. This capability makes it especially useful in imaging the brain, the muscles, the heart, and cancers compared with other medical imaging techniques such as CAT scans or x-rays (both of which I cover earlier in the chapter).

Another difference between MRI and x-ray is that MRI focuses on the hydrogen atoms within water. The human body is approximately 70 percent water, and the MRI can differentiate between fluid regions and solid areas such as organs, tumors, or bones. For example, check out the MRI image in Figure 17-4 of the knee.

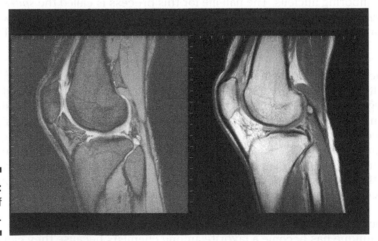

Figure 17-4: MRI scan of the knee.

© Nick Veasey/Getty Images

Chapter 18

Optics Everywhere: Exploring Other Medical, Industrial, and Military Uses

. . . .

In This Chapter

▶ Removing tissue and promoting healing with lasers

▶ Exploring a wide variety of industrial and commercial applications of optics

▶ Understanding how the armed forces and law enforcement utilize optical devices

. . . .

*I*maging has been a primary use of optics for many centuries, but with the advent of the laser and fiber optics, many other applications have come into play. (For more on the basics of lasers and fiber optics, check out Chapters 14 and 15, respectively.) In this chapter, I describe some of the other uses for optics to give you an idea of all of the subtle places that a knowledge of optics plays a role, including applications in such diverse fields as medicine, commerce and industry, and military and law enforcement applications.

Considering Typical Medical Procedures Involving Lasers

Medical procedures to look into the body (such as the optical processes in Chapter 17) often find materials or tissues that need to be removed or holes that need to be plugged. Although traditional surgical procedures are still very common, more surgeons are turning to lasers as their instrument of choice.

I wish laser procedures were as simple as pointing a laser at the problem spot and zapping it to make it better, but much more study is necessary to figure out how to use lasers properly. Optical engineers working with doctors must understand the optical properties of the environment that they want the device to work in and then select the best configuration for the laser to be used. After they find an optimized procedure and device, they know lasers typically offer fewer complications and better results than scalpels in certain circumstances. The following sections look at two common types of laser procedure: ablation and suturing.

Removing stuff you don't want: Tissue ablation

The earliest use of lasers in medicine has been to get rid of stuff; a more official-sounding term for this process is *ablation*. Most medical applications use lasers to burn tissue. The laser's energy is high enough that it heats up the area, causing the tissue to break down, usually into simpler things such as atoms or simple molecules.

Mythbusting time: Although lasers in the movies can blow stuff up in a rather spectacular fashion, current real-world lasers don't work that way. Lasers do technically use heat to eliminate material, but the material in question very seldom explodes, unless you shoot something that's already flammable. So, tissue ablation is no more exciting to watch than seeing a little flash of light and a puff of steam, which is why movie directors probably decide to enhance laser scenes a little.

Of course, you have to account for some special factors when designing medical laser systems for ablation, and no one laser and laser delivery system is right for every application. The following sections present the main considerations involved in laser ablation systems.

An actual laser ablation procedure is much more complicated than what I describe in this section, but this discussion gives you an idea about how much optics this often-desirable procedure involves.

Selecting the proper laser

To choose the proper laser source for a particular ablation procedure, the doctor determines the necessary pulse energy and wavelength.

Most (but not all) medical lasers are pulsed lasers — that is, the laser light is on for relatively short amounts of time, like several hundred picoseconds or tens of nanoseconds. Pulsed lasers provide three advantages:

✔ **They allow for more-efficient extraction of energy from the laser cavity.** More-efficient laser energy extraction means that you can run lasers with lower electrical power, which is always a plus. Besides energy savings, this efficiency also reduces the amount of cooling required to keep the laser operating safely.

✓ **They have a nearly constant amount of energy per pulse.** This characteristic is important to allow the doctor precise control over how much tissue is removed. A burst of energy that is the same amount each time allows the doctor to remove small, predictable amounts of tissue, which is really important when you operate near healthy tissue.

✓ **They're more easily controlled (turned on and off at exactly the correct time).**

The last two characteristics are important to reduce collateral damage — that is, laser energy going somewhere that it wasn't supposed to go and damaging healthy tissue. Sometimes, some healthy tissue needs to be removed, especially when dealing with cancer. But having the doctor in control of exactly how much surrounding tissue is removed is much better than having an indiscriminant amount removed.

Body tissues are generally red because of the blood, so they usually readily absorb anything with a green or blue wavelength. Selecting the wavelength of the laser is a little more challenging. Most medical lasers these days emit light in the infrared region, something with a wavelength greater than 700 nanometers. The lasers that work in this wavelength region (near infrared region) are much more efficient than lasers that work in the visible region. Many body tissues don't readily absorb this type of light. To get around this problem, optical engineers either find a wavelength that the tissue more readily absorbs or increase the power of the laser light, which doesn't help with efficiency but does make it work by making the small percentage of the incident light absorbed be a larger total energy so that it can ablate material.

Another, increasingly common tool is a nonlinear optical process called *frequency conversion,* which I talk a little more about in Chapter 20. Frequency conversion happens in certain crystals where the infrared light is converted into higher frequency light when nonlinear interactions occur with the atoms or molecules in the crystal. To create these lasers, designers place nonlinear crystals in infrared lasers so that the light the lasers emit is in the green or blue region of the spectrum. This modification means that doctors can use lower laser powers because the light is much more effectively absorbed.

Delivering the light on target: Fiber optics

Getting the laser light to the targeted area usually involves fiber optics. Aside from some dental and eye procedures, which can use a *free-space* laser beam (a beam directed by mirrors), delivering the laser without opening the body cavity is usually more convenient, and it's a minimally invasive procedure that can minimize many other risks associated with other methods of tissue or material removal from the body. During the actual procedure, the fiber-optic cable is sent either through an artery (through a catheter) with x-ray guidance or through a small incision made into the body (arthroscopically).

Using fiber optics to direct a medical laser has some challenges:

✔ **Spectral dispersion:** As I note in Chapter 15, *spectral dispersion* occurs when long-wavelength spectral components of the laser pulse get to the end of the fiber before shorter-wavelength components, making the pulse longer than it was when it came out of the laser cavity. Spectral dispersion affects some laser applications more significantly than others. If a surgical application requires a *peak pulse power* (minimum amount of energy delivered in a certain, small amount of time) in order to break apart some tissue to the molecular level, spectral dispersion can make that procedure much less effective or even ineffective. You have to account for the effects of spectral dispersion in all medical lasers, if only to determine whether they're significant.

✔ **Melted fibers from high peak powers:** Fibers with extremely small cores (such as those used to combat intermodal dispersion — see Chapter 15) often wind up melting in medical laser applications. The smaller the core, the less energy the fiber can handle; this small core is more problematic with pulsed lasers because the peak power is already larger. Add in even more peak power caused by focusing the light to a small beam diameter, and you often end up with a melted fiber.

To avoid fiber meltdown, medical lasers typically use larger fiber cores. This solution increases the effect of intermodal dispersion, but using a focusing optic, such as a small lens, at the end of the fiber can often compensate. The focusing optic doesn't shorten the pulse, but it can increase the peak power by reducing the beam size at the laser focal point. (*Note:* Although intermodal dispersion is a problem over kilometers of fiber, it doesn't significantly affect the meters of fiber typically used in laser medical applications.)

Another way to deal with high peak powers melting fibers is to use a *pulse stretcher* (or just *stretcher*), a device that (surprise!) stretches the pulse out before sending it into the fiber. This device spreads the laser's energy out more in time, significantly dropping the peak power so it can go into a small fiber safely. Near the end of the fiber, the stretched pulses are sent through a *pulse compressor* (or *compressor*), which compresses the pulse's energy in time and drives the peak energy back up so that, when coupled with the focusing optic, the laser light has the desired peak power for the particular operation.

Knowing the optical properties of the target: What does the laser do to the tissue?

An important aspect of laser ablation is understanding how the tissue will work with the laser. Obviously, if the material you're ablating is more likely to reflect the light than absorb it or the surrounding tissue is much more sensitive to the laser light than the tissue you want to remove, going with the laser may be too risky.

If you decide that the laser is a viable option, you need calculate how much energy you need to send into the material to remove it, including considering how much laser light must be used to remove the material completely without unnecessary damage to surrounding desirable tissue.

Lasers: Preventing trauma and keeping you from bumping into walls

Using the laser rather than a scalpel to remove material reduces the risk of strokes and heart attacks. Strokes can occur when a large enough amount of material effectively blocks a blood vessel and prevents blood flow to the brain; if the material blocks a blood vessel in the heart, it causes a heart attack. Laser ablation typically dissociates the material into very tiny particles, smaller than what can plug the small arteries in the brain or heart.

You may be familiar with another application of laser ablation: LASIK surgery, which changes the curvature of the cornea to correct for common focusing problems (nearsightedness, farsightedness, and astigmatism; see Chapter 13 for more on these conditions). The basic idea for LASIK is to cut a small flap out of the upper surface of the cornea (with either a laser or a special scalpel) and then use a laser to ablate material from very precise locations in the lower cornea material. This process changes the shape of the cornea when the flap is laid back down and heals.

Sealing up holes or incisions

Not only do lasers remove unwanted materials from the body (see the earlier section "Removing stuff you don't want: Tissue ablation"), but they can also close up holes or incisions. In fact, one of the benefits of using lasers for ablation is that they can cauterize surrounding tissue at the same time, which reduces some of the risk associated with surgeries. Technology isn't at the point where you can just wave a laser over a paper cut and make it go away (if only!), but it's getting closer. In the following sections, I look at some common ways lasers can help medically seal holes or incisions.

Suturing

Suturing (closing up incisions or sealing holes) traditionally involves staples or special thread (known as *sutures*).The body may dissolve the sutures after some time when the cut or hole heals, potentially leaving a noticeable scar you may or may not want. A faster, less-scarring way to suture involves lasers. In this application, the laser changes the material's structure so that the remaining material aids the healing process.

Because you can't just melt the skin to close a cut, you need another material that functions like a solder would for joining two pieces of wire; researchers are trying various types of materials. Laser suturing also means that you don't have to worry about having sutures come out prematurely; the laser suture usually is stronger than the nonlaser sutures, primarily because laser suturing is continuous (along the entire cut). The only current problem with laser suturing is that it only works on small holes or smooth incisions. A gash in the skin still requires traditional sutures, but research is continuing. In fact, researchers are looking to see if "tissue welding" is possible for larger cuts or holes requiring additional material, the tissue equivalent of the welding rod in metal welding. This technology may be useful for repairing tears or holes in many other structures in the body, like arteries.

Fixing retinal problems

Lasers can also seal a perforation in the blood supply in the retina, a problem that can damage the healthy part of the retina and potentially cause sight loss. Shining a laser through the eye and focusing it on the leaky artery works to cauterize the hole, stopping the flow of blood into the retina and eyeball.

Retinal tears are another procedure that can be handled with lasers. *Retinal tears* usually occur in conjunction with some kind of head trauma, like hitting your head on the steering wheel during a car accident. Part of the retina detaches from the eyeball and rises above it, obscuring the view in portions of the intact retina. Traditional, invasive surgery can repair some types of retinal tearing but runs the risk of serious side effects, including loss of sight in that eye. If the tear is sufficiently small, doctors can use a pulsed laser to tack the loose part back in place. (If the tear is larger, something other than the laser is required to get the retina back into its original position before securing it back in place.) As long as the nerves and blood supply aren't damaged, sight can be regained in most of the reattached part.

Purely cosmetic: Doing away with tattoos, varicose veins, and unwanted hair

Medical lasers can change the appearance of certain features on the body. One of the biggest cosmetic laser uses is tattoo removal. Tattoos are designed to be permanent, so they're often very difficult to remove. (I *am* talking about the permanent ones here; I don't recommend removing stick-on drugstore tattoos with a laser.) Aside from scraping the stained material away, a properly designed laser system can focus the laser on the ink dots and dissociate the ink, often making it disappear. This procedure can still result in slight scaring that may retain the shape of the tattoo, but it generally looks better than having the tattoo scraped off (or, in some cases, keeping it altogether).

Lasers are also good for removing *varicose veins,* those blue lines that appear very close to the surface of the skin (they're blue because the blood isn't oxygenated and being returned to the heart). Varicose veins on your face, arms, or legs can make you feel unattractive; you can deal with them surgically, but a laser can cauterize and dissociate the remaining material with minimal side effects.

Another popular cosmetic laser procedure is laser hair removal, which produces minimal scarring and produces rather long-lasting results. The technician chooses a laser wavelength such that the laser damages hair follicles because they absorb more light than the surrounding, lighter skin does. This characteristic means that the procedure works best on people with dark hair and light skin, who are usually the folks that want this procedure done.

One of the primary advantages of using the laser in cosmetic applications is that the procedures can be done as an outpatient, thereby reducing the cost of the procedures. Certainly not all procedures, but many more than would be possible otherwise without the knowledge and availability of lasers and optics.

Getting Industrial: Making and Checking Products Out with Optics

Optics has played a major role in many different applications related to production. Traditionally, optics's main job has been to verify the quality of items made or materials provided, but with the advent of lasers, many more applications have become part of the actual fabrication process or the function of the final product. In this section, I highlight some of these applications.

Monitoring quality control

Manufacturers can use interference and holography to make sure items are flat or properly curved. To within the size of the wavelength of light, optical interference arrangements can measure the heights of structures or the flatness of an area. Interference can also test high mechanical stress parts, like aircraft or jet engine turbine parts.

Holography (see Chapter 16) is helpful in identifying problems in multiple areas of manufacturing, including the following:

- Changes in structure caused by heat or mechanical stress.

- Metal fatigue or *microfractures* (very tiny cracks in surfaces). This capability can help engineers either figure out why parts failed or notice weak parts before they result in a catastrophic failure.

- Parts made incorrectly or with materials that have structural defects in them as a result of something in the manufacturing process.

Drilling holes or etching materials

Lasers can drill holes in a wide variety of materials, from metals to semiconductors, without the need for sharpening or the drop-off in performance quality you often get with drill bits. This advancement has provided some much-improved manufacturing output and improved efficiency in the operation of devices for the semiconductor, medical devices, engine manufacturing industries, and many others.

Drilling large holes with a laser isn't practical, but lasers are very effective for tiny holes or tiny structures.

If finer detail work is what you're after, lasers are great for ablating material along a particular pattern. Whether large or small, a laser can make intricate patterns relatively easily and often more effectively than milling or even wet etching can.

For etching, the laser beam usually has to undergo conditioning to produce a top-hat irradiance distribution as opposed to the usual Gaussian distribution. This setup takes a lot of design work, but a properly built laser can not only etch squared channels but also clear out areas with a flat bottom.

Making life easier: Commercial applications

Many chapters in this book discuss commercial applications, especially related to entertainment, including projectors, cameras, 3-D movies, and fiber-optic communication. One I don't cover very extensively is one you're probably familiar with: the data storage devices called DVD or Blu-ray Discs. These devices store digital information on discs that reflect different amounts of light, producing a signal like an electrical binary stream of ones and zeros. Unlike their magnetic counterparts (such as hard-disk drives), magnetic fields don't affect the data.

However, you've probably noticed one of the liabilities of this optical technology: Because the information is transferred to your video device based on the reflection of a laser off the disc's surface, any significant scratches or residues from fingerprints lock up the player. So take care with your Blu-ray Discs and DVDs; with great capacity comes great responsibility.

If you're a person who likes to do some projects around the house (maybe you needed some shelves for all those video discs), you may have used another commercial optical device: a laser level. This handy device spreads a horizontal or vertical line made by laser light (no, it doesn't burn a line in the wall). The device has internal sensors that orient the beam correctly for the orientation you select. With the press of a button, the small diode laser inside sends light through a series of optics that project the beam in a thin line so that you can hang your picture correctly or line up your door frame vertically.

Applying Optics in Military and Law Enforcement Endeavors

Two important fields where optics is extremely helpful are military and law enforcement. Whether they're gathering information with the help of visible and invisible light, working in low-light situations, or blowing up targets with energy carried by light, professionals in these fields use some of the following optical applications to keep people safe at home and abroad.

Range finders

One of the early optical technologies on the battlefield was range finding. Range finders now are fairly standard pieces of equipment on many different missions. They've helped change battlefield tactics quite a bit and have played a major role in reducing collateral damage in many limited-engagement actions. Before present range finders, officers had to estimate the range to targets from observations. This method made moving targets very difficult to hit and increased the likelihood of collateral damage. To reduce this undesirable effect, the laser range finder is able to determine precisely the range to the target and take readings of the target position quickly; the range finder can then calculate the target's speed, and computer software can plot a rather accurate trajectory to the target.

Of course, the farther away the target is, the more difficult it is to target accurately because of atmospheric effects. *Atmospheric effects* manifest themselves as variations in the index of refraction of the air. Because the air's index of refraction is nominally proportional to its density (see Chapter 4), as the air temperature increases, the air's density and index of refraction also decrease, and vice versa. As the laser travels kilometers through the air, the fluctuations in the air's index of refraction can make the laser beam wander. Luckily, processes can correct for this atmospheric refraction effect.

These atmospheric effects are what cause the stars you see from your ground-based telescope to twinkle. From the space station, the stars don't twinkle because there's no air and therefore no index of refraction to change.

Range finding has its limits. You can only deliver an ordinance so far from the ground, and targets on the ground can often be camouflaged so that they aren't easily identifiable from the air. (That's where target designation comes in; see the following section.)

Related to the range finder is a piece of equipment that may have been used against you: the laser velocimeter or laser speed gun. Although the instruments for catching speeders have traditionally been radar guns, some police departments now have laser-based devices (superior to the radar speed gun) that can measure your speed, often without your being able to detect it. Radar guns are easily detected and jammed (hence the rather large market in radar detectors and jammers) because the beam they emit is very broad and on for a long time. Laser speed guns better avoid these evasive devices because the laser beam is very small (practically a small dot), and the pulses are very short; however, these devices also cost more than radar guns.

Target designation

Target designation is another laser based advance in battlefield techniques. It involves a laser pointed at a target from either the ground or the air. The laser light scatters off the target (called *painting the target;* head to Chapter 4 for more on scattering). Then an aircraft (another aircraft, if one also painted the target), usually at much higher altitude, drops a *smart bomb* that has a special detector on the front that can control guide fins at back. The detector looks for the scattered laser light and guides the bomb to the bright spot, the target.

Along with range finders, target designators and their smart bomb counterparts have significantly reduced collateral damage, even in city targets. In the old days, many parts of a city had to be destroyed in order to take out a particular target of military value. With smart bombs and target designators, only the target is destroyed.

As with all laser propagation through the air, each new system needs a lot of work to make sure it accounts for all the factors that can influence where the laser beam goes, including verifying that the optics in the designator system are working properly. Missed targets (such as occurred in some of the Kosovo campaigns in the 1990s) are probably due to a misalignment of the optical system in the target designator system. The systems are relatively complex, and the military environment isn't exactly compatible with precision optical systems. However, they still have a very impressive track record.

Missile defense

One optical technology that's often in the news these days is missile defense. Most of the short-term missile defense solutions are antimissile missiles, which can be sea-based (with a navy) or land-based (with an army). These options are usually designed to take out short- to midrange rockets or missiles. The (very big) challenge is to get these devices to hit the incoming missile quickly, so some researchers have begun to look at using lasers to take out missiles. The advantage of the laser is that it travels at such a high rate of speed that it's practically instantaneous. The researchers have squared away pointing and tracking issues; the remaining problem is that the laser beam needs to be very powerful to melt a critical system on the incoming missile in a short amount of time, and sending a large amount of power through the atmosphere is more easily said than done.

High power laser beams heat up the atmosphere significantly, leading to a phenomenon called *thermal blooming* that can cause the beam to lose significant amounts of laser energy, reducing the laser power (often significantly) when the beam hits the target. The *beam director,* the optical system used to point the laser beam at the target, not only keeps the laser on target but also delivers the energy in such a way as to minimize blooming. Joint programs between the United States and Israel have demonstrated that high powered lasers can adequately take out medium-range Katyusha rockets.

The Airborne Laser (ABL) is a very powerful U.S. Air Force laser system based in a large aircraft. This system targets long range intercontinental ballistic missiles (ICBMs) in their boost phase, which is the best time to destroy them all in one place. Most ICBMs have decoys in them to defeat missile countermeasures, so stopping an ICBM attack is difficult, at best, if the warhead deploys. The ABL is powerful enough to hit critical systems in the missile, causing it to blow up before it can deploy the warheads and the decoys.

Night vision systems

Night vision systems are arrangements that help you see in low- and no-light situations, which both military and law enforcement personnel encounter. The two types of night vision systems are image intensifiers and near-infrared vision:

- **Image intensifiers:** *Image intensifiers* (known as I^2, pronounced "I squared") are devices that work with low visible light levels, such as nighttime with a clear sky. This situation has lots of light, just not enough for you to see things with enough detail to be helpful. I^2 devices capture the light and then amplify it before sending it to your eyes in the infamous green color. I^2 devices are used by all branches of the military and many law enforcement groups when stealth is an advantage.

- **Near-infrared:** *Near-infrared night vision systems* (traditional night vision) use the near-infrared light that's much more common at night than visible light is. Night vision systems have a special device that converts the infrared light to green light that you can see. (It uses green light because your eye is sensitive to that light, which doesn't require large amounts of energy to function.)

  Some night vision systems can use with longer-wavelength light (closer to the midinfrared range) so that infrared flashlights can be used without detection by older night vision systems or the eye.

Thermal vision systems

Thermal vision systems look at relative temperature differences between different areas. Because these systems show temperature differences rather than reflected light, they don't make images that you're used to seeing. Instead, areas at the same temperature are the same color, so a temperature change from one spot to another is indicated in the image as a different color.

These devices can track individuals by looking at their thermal footprints on the floor or ground. Thermal vision is also effective in finding people, vehicles, or even illegal drug labs that are camouflaged against visible light. Law enforcement can look at a row of cars and tell which ones were running recently, as well as make a reasonable guess as to how long the motor has been off or, based on the heat signature on the tires, how long of a trip a vehicle made.

Image processing

Although you may be used to seeing detective shows using image processing to get license plate numbers off vehicles from security or traffic cameras, image processing actually helps gather all kinds of information in various fields.

Image processing is the name given to computer programs that can perform mathematical calculations on the patterns of light and dark in a relatively low-resolution image and generate fine details. Images of far-away objects or from satellites in orbit are generally pretty fuzzy and seemingly useless. But combining the science of diffraction and mathematic pattern recognition, image processors can enhance image details, such as the general features on a face or the type of vehicle in the image, which is especially helpful in law enforcement. Granted, actual image processing usually doesn't happen as quickly as it does on TV (what does?), but it's still a viable law enforcement tool for cracking cases.

Although the military and law enforcement have particular uses for image processing, those applications are just the tip of the iceberg. Image processing can give you information about urban and suburban areas, crop or vegetation information, and even rock formation differences and water density information. The type of detail that you can get depends on the type of image taken (such as whether it's from a satellite or security camera) and the differences you need to see to identify things of interest to you (such as mineral concentrations or vegetation types).

Chapter 19

Astronomical Applications: Using Telescopes

Since ancient times, humans have looked into the heavens with awe, but only relatively recently have we figured out how to develop tools that help us look more closely at the distant objects that have always fascinated us. These tools are called telescopes, and in this chapter, you find out how they work.

Telescopes come in a wide range of designs, and each works best for different purposes. All of them work by taking the light from a distant source — a faraway building or a star, planet, nebula, or other cool astronomical body — and focusing it in a way that allows the eye to clearly resolve a bigger, brighter image than it could otherwise.

Note: This chapter focuses on the development and evolution of the telescope instead of actually providing a guide to using them. If it inspires you to take to watching the night sky yourself, you'll certainly want a lot more practical advice than what I offer here. A more how-to approach to using telescopes is laid out in *Astronomy For Dummies,* 2nd Edition, by Stephen P. Maran, PhD (Wiley).

Understanding the Anatomy of a Telescope

The purpose of a telescope is relatively straightforward: Focus the rays of light coming from a distant object into an image that's more clearly viewable close up. When pointing a telescope toward the heavens, the goal is to make the image both larger and brighter than what you can see by looking into the sky with the naked eye.

Simple enough, you may think, but putting all the pieces together to achieve this goal took a while. After it was done, however, the technology spread fairly quickly, maintaining the same basic design shown in Figure 19-1.

Figure 19-1:
A refracting telescope uses lenses to collect and focus light.

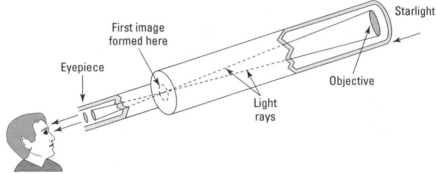

First image formed here

Eyepiece

Starlight

Objective

Light rays

Since their creation in 1608, most telescope designs (whether they use lenses, mirrors, or both) have included the same basic elements:

- ✔ The telescope gathers the light.

- ✔ The image is viewed through an eyepiece (or, in more modern telescopes, recorded and analyzed using photographic film, digital cameras, or spectrographs).

Each of these elements is crucial to how the telescope functions, though the process of how the light gets magnified into the final image, and how that final image is viewed, is different for each design. Some models may contain bells and whistles along the way, but the basic concept is the same.

Gathering the light

The light comes into a refracting telescope through the *objective* (sometimes called the *primary lens*). The diameter of the objective is called the *aperture*. In other words, the bigger the objective the more light the telescope can gather and the better the viewer should be able to study distant objects. Light-gathering power is especially helpful in studying galaxies and quasars, the most distant objects in the universe.

For reflecting telescopes, the same concept applies, but it's an objective mirror instead of an objective lens. See the section "Reimagining Telescope Design: Reflecting Telescopes" later in this chapter for more about how these concepts show up in reflecting telescopes.

Viewing the image with an eyepiece

Those of us who aren't astrophysicists (or other space professionals) probably view the image through an *eyepiece,* which is a lens (or series of lenses, in modern telescopes) that bends the image back into a form that your eye can resolve clearly.

Today, most eyepieces are actually a series of lenses mounted together in a metal or plastic housing. The lens closest to the eye is called the *eye lens.* (Creative name, I know, but astronomers are not known for creativity when naming telescope designs or components.)

The eyepiece lens farthest from the eye is called the *field lens.* The purpose of the field lens is to expand the useful *field of view* (how much of the scene you can see at a given time) when looking through the telescope. This multi-lens eyepiece design was developed in the mid-1600s, less than 50 years after the appearance of the first telescope.

The field of view means how much of the night sky you can see at a given time. If you want a high magnification in order to study the fine details of an object, you reduce the field of view and can therefore see a smaller portion of the sky at one time. Essentially, swapping out an eyepiece gives you a different observational experience depending on what you're trying to achieve.

One way to think of the telescope is that the objective forms a small image of a very big object that's far away. The eyepiece (possibly along with other lenses) then takes the very small image and magnifies it again before sending the image into your eye. (Check out Figure 19-2 for an illustration of this progression.) You need both parts to have a telescope.

Note: Telescopes used by professional astronomers doing research do not actually incorporate eyepieces, because they aren't actually viewed with the human eye. Instead, a digital camera or other device records the light coming through the telescope.

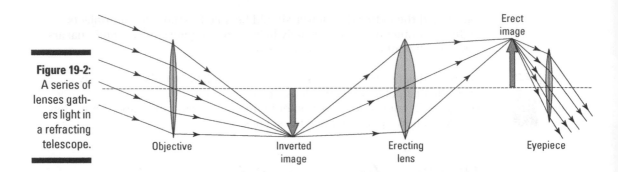

Figure 19-2: A series of lenses gathers light in a refracting telescope.

Objective   Inverted image   Erecting lens   Erect image   Eyepiece

Revolutionizing Refracting Telescopes

After glassworkers began making lenses and then started using them to make spectacles, the discovery of the telescope was a virtual inevitability. *Refractors*, or refracting telescopes, were the first type developed. Chapter 13 discusses the science behind this most basic form of telescope.

The first confirmed refractor was a spyglass presented by spectacle maker Hans Lipperhey to Prince Maurice of Nassau (in the Netherlands, not the Bahamas) in 1608. This early device probably had a magnification of about three times, meaning it could view an area three miles away as if it were only one mile away. This achievement may not seem incredibly impressive, but it was a potentially useful military application; that made the technology instantly of interest to Prince Maurice, who was in the middle of peace negotiations with the Spanish. (Why Spain was fighting the Netherlands, I have no clue, nor is it relevant to the story.)

Maurice formed a committee to investigate the matter, and it asked Lipperhey to work on a design for a device that would work with both eyes. Though this early attempt at binoculars didn't bear any particular fruit, it showed the interest in the strange new technology and the desire to see what could be done with it.

At around the same time, Prince Maurice showed off his new device to the Marquis Spinola of Spain, the very military adversary who was in town for the negotiations! Spinola returned to Spain, by way of Brussels, telling his allies along the way about the curious invention he had seen.

The fact that Maurice was so bad at keeping this particular military secret may be historically significant. Word of the telescope certainly seems to have spread quickly across Europe, reaching an Italian mathematician, Galileo Galilei, in May 1609. The following sections look at some advancements to the basic refractor design after this point.

Discovering the telescope: Great minds think alike

The birth of the telescope is generally traced to Hans Lipperhey in 1608 Netherlands, but evidence suggests that telescopes were simultaneously springing up in Frankfurt, several hundred kilometers away. Still other claims place similar devices in Italy around 1604, while some information hints at even more-ancient designs.

Around 1570, Englishman Thomas Digges's book *Pantometria* described a device that appears to be some sort of reflecting telescope, claimed to be created by his father, Leonard Digges (1520–1559). Digges claimed it could see seven miles, which most scientists dismiss. A century later, Newton's reflector didn't have that kind of range, so Digges's technology probably couldn't have achieved it.

Still, this account is corroborated by mathematician William Bourne, who himself proposed a design for a one-lens device similar to a telescope. The problem is that this light would enter the eye converging, not parallel, resulting in a blurry image.

These ideas by Digges and Bourne are not completely unprecedented. In his book *Opus maius,* 13th-century Franciscan friar Richard Bacon suggested arranging lenses in "such a way with respect to our sight and objects of vision, that the rays will be refracted and bent in any direction that we may desire, and under any angle we wish we shall see the object near or at a distance." Bacon also claimed that Julius Caesar had used a spyglass of this sort of design, suggesting the Romans had stumbled upon rudimentary telescope technology. (But no real historical evidence supports Bacon's claim.)

Galilean telescope

The basic refracting telescope is often called a *Galilean telescope,* even though Galileo didn't invent it. The key principles of refraction were already known, but Galileo was the guy who analyzed, improved upon, and, more importantly, popularized the telescope as a scientific tool.

The Galilean telescope consists of two lenses, a convex objective and a concave eyepiece. The objective takes the light coming in and focuses it into an image beyond the eyepiece, which then bends the light back so that it enters the human eye as parallel light rays, as shown in Figure 19-3. The effect is that the image projected on the eye appears larger.

The basic Galilean telescope design allowed Galileo to make several groundbreaking discoveries, such as finding the moons of Jupiter, the feat that formed the basis of his 1610 book *Starry Messenger.*

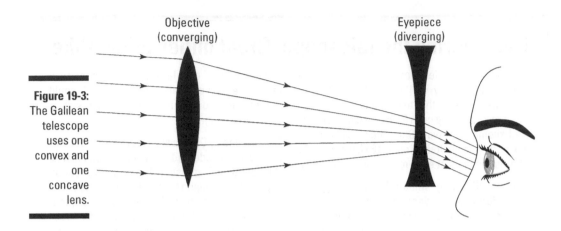

Objective
(converging)

Eyepiece
(diverging)

Figure 19-3:
The Galilean
telescope
uses one
convex and
one
concave
lens.

Kepler's enhancement

Astronomer Johannes Kepler quickly learned of Galileo's publication and became instantly fascinated. He began an exhaustive analysis of the optics of telescopes, culminating in his 1611 book *Dioptrice*. In it, he proposed a new design called the *Keplerian telescope*.

The Galilean telescope had one major drawback: the better the magnification, the worse the field of view. In a sense, using Galileo's telescope was like looking through a long drinking straw. The better you saw the thing you were looking at, the more restricted you were from seeing anything else.

Kepler realized that this flaw was due to the concave eyepiece, so he proposed swapping that lens out for a convex one. (Being more of a theorist than an experimenter, he never actually built a working prototype of his design.)

One of the major benefits of Kepler's design was that the magnification was extremely easy to calculate. The magnification was just the objective's focal length divided by the eyepiece's focal length.

Kepler's design had a drawback as well, and a fairly major one: The image that came out of the telescope was inverted. This design would be horrible for a military commander trying use a telescope to spy behind enemy lines, but it wasn't a big problem for astronomers. Astronomers wanted to see astronomical objects; getting a closer look at Jupiter upside-down is, after all, still getting a closer look at Jupiter! For this reason, the Keplerian telescope is sometimes called an *astronomical telescope* or *inverting telescope*.

In *Dioptrice*, Kepler realized this drawback and offered a solution: If you introduce yet another convex lens, called an *erecting lens*, that lens inverts the image a second time so that the final image is upright. (Figure 19-2 earlier in the chapter shows this design.) However, this setup created another drawback because it resulted in much longer telescopes. The telescope had to be increased by a size of about four times the focal length of the erecting lens.

Tycho's keen eye

Johannes Kepler had very poor eyesight, so he may never have actually used a telescope. Instead, Kepler's work was based on the notes of his mentor, Tycho Brahe, who died in 1601 without ever having the benefit of a telescope.

Brahe did, however, travel Europe and take measurements at several of the most advanced observatories, which had devices that were able to help pinpoint positions in the skies without the benefit of telescopic magnification.

Most credit Brahe with the most extensive and precise astronomical observations of any pre-Galilean astronomer.

In fact, Kepler was able to use Brahe's records to develop and publish his first two laws of planetary motion in 1609, before he ever heard of Galileo's *Starry Messenger.* (Kepler's third law was published in 1619, and by that point he may well have been using data gathered with telescopes.)

Reimagining Telescope Design: Reflecting Telescopes

Within a few decades of the refractor's invention, designs sprang up across Europe for telescopes that used mirrors rather than lenses. Two major proponents of this idea in the early- and mid-1600s were René Descartes and Minorite friar Marin Mersenne.

These *reflectors,* or reflecting telescopes, had the advantage of creating less distortion than refracting telescopes did, but the technology took a few years to catch up with the theory.

Newtonian

The first successful reflector prototype is generally attributed to the 1671 telescope built by Sir Isaac Newton. This *Newtonian telescope* design used mirrors to reflect the light through an eyepiece on the side, magnifying the image along the way, as shown in Figure 19-4.

The light enters through the top of the telescope and travels its length, being collected by the large, concave parabolic *objective mirror* at the base. On its way toward forming an image back up the length of the telescope, the light collides with a plane mirror that's placed at an angle. This mirror is called the *diagonal* or *flat.*

The diagonal's job is to reflect the light through the eyepiece, which is mounted perpendicular to the telescope. The final image ends up being formed right on the viewer's retina.

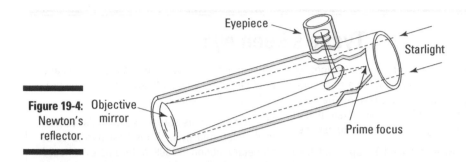

Figure 19-4: Newton's reflector.

This design has the benefit of convenience, among other things. The eyepiece is near the top of the telescope; if you've ever tried to look through the base of a refractor, you know that it can sometimes be a pain to crouch down, especially if you have joint or lower-back pain. (See the sidebar "Reflector versus refractor" for more on how the benefits of each type of telescope stack up.)

Cassegrain

The *Cassegrain telescope* (see Figure 9-5) uses a concave objective mirror as well, but instead of reflecting light out the side, it reflects the light off a convex mirror called a *secondary*. The secondary reflects the light back toward a hole in the base, right in the middle of the objective.

The telescope gets its name from the inventor, Monsieur Cassegrain of Chartres, who announced the design in 1672. It couldn't be built at the time, however, because the eyepiece lens had to be hyperbolic . . . a shape that couldn't actually be created with existing technology.

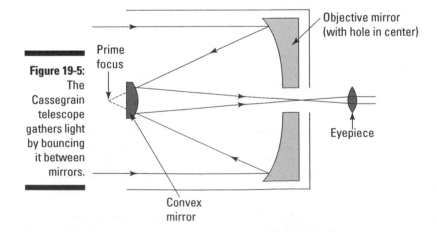

Figure 19-5: The Cassegrain telescope gathers light by bouncing it between mirrors.

Today, the Cassegrain design is quite common, though not so much with amateurs. Large telescopes used for professional work are often based on some variant of the Cassegrain design.

Gregorian

Years before Cassegrain (see the preceding section), Scottish mathematician James Gregory struck on a design, called the *Gregorian telescope*, that was very similar to the Cassegrain design except that it used a concave secondary rather than a convex one. You can see a Gregorian telescope in Figure 19-6.

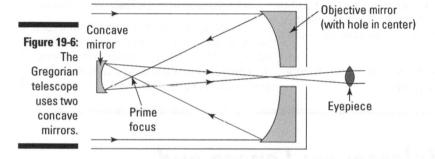

Figure 19-6: The Gregorian telescope uses two concave mirrors.

At the time, some scientists argued that the Gregorian design was the first reflector, because Gregory began working on a prototype in 1662 before getting distracted by a European trip. This work was later picked up by Robert Hooke, but the tests were ultimately unsuccessful, possibly because the mirrors weren't yet high enough quality.

Today, Newton gets credit for the first reflector because he was the first to build one successfully in practice and not just on paper, but he, Gregory, and Cassegrain had spirited debates through correspondence (the 17th century equivalent of an Internet flame war) with the Royal Society about who should rightfully be credited with creating the first reflecting telescope.

Reflector versus refractor

Both refracting and reflecting telescopes have their fans, and for good reasons. No one type of telescope is better at everything. Here are a few of the main advantages and disadvantages of each:

✔ Because lenses refract light like a prism, refracting telescopes can create halos of color around objects. Some techniques use different lens materials to correct for this problem, creating a nearly *achromatic* ("without color") image, but the mirrors in reflecting telescopes avoid this problem altogether.

✔ Refractors are much bulkier than reflectors of comparable quality.

✔ Optical errors on a reflector's mirror cause about four times the effect on final image quality as the same level of optical error on a refractor's lens.

✔ Although a mirror can get somewhat better resolution, it's also more sensitive to adjustment errors than a lens. If your adjustment is slightly off with a refractor, you can still get a good image. If it's slightly off with a reflector, your image is likely not very good.

Overall, refractors seem to be a bit more forgiving of errors (human and otherwise) than reflectors. For beginning amateur astronomers, this characteristic alone may be enough reason to start out with a refractor.

Hybrid Telescopes: Lenses and Mirrors Working Together

Many modern telescopes use a combination of refraction and reflection in an attempt to maximize their effectiveness. Such hybrid systems are called *catadioptric telescopes,* and I discuss some common ones in this section.

Schmidt

One intriguing hybrid design is the *Schmidt telescope*, designed in 1930 by Bernhard Voldemar Schmidt. In this design (shown in Figure 19-7), a lens and mirror work together to focus light on a small plate inside the telescope rather than through the eyepiece. The plate contains a *photographic emulsion,* or light-sensitive surface, to record the image.

In stark contrast to other telescopes earlier in the chapter, the Schmidt contains a lens called a *corrective plate* at the front end of the telescope. (The curvature of the corrective plate is exaggerated in Figure 19-7 for effect.) This plate was designed to help correct for optical problems that occurred when trying to take photographs of a nebula.

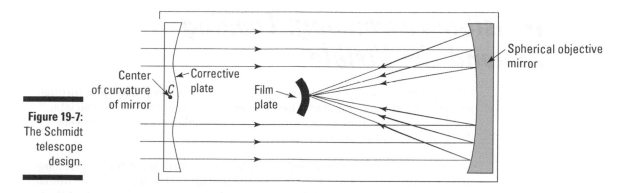

Figure 19-7:
The Schmidt telescope design.

Labels in figure: Center of curvature of mirror · C · Corrective plate · Film plate · Spherical objective mirror

An optical aberration known as *coma* made stars look like little comets in photographs. Stars in the center of the image looked fine, but as you moved toward the edges, it got worse. With parabolic mirrors, this effect is noticeable within a fraction of a degree from the center of the field of view. Schmidt threw out the parabolic objective mirror and replaced it with a spherical one. The problem was that a spherical objective mirror created spherical aberration, so Schmidt had to correct for that, too. The corrective plate does just that. These two adjustments together had the effect of eliminating coma and creating a good wide-field telescope.

Some telescope designs, called *Schmidt-Cassegrain telescopes*, basically follow the Schmidt design but have a convex mirror rather than a photo plate. The light is then reflected out of the telescope through a hole in the objective, as in the Cassegrain design (see the earlier section "Cassegrain" for more on that design).

Maksutov

In 1941, Russian optician Dmitri Maksutov developed a variant of the Schmidt telescope called the *Maksutov telescope*. (Bet you didn't see that name coming!) The major difference in this design is that instead of using a basically flat but slightly wavy corrective plate, Maksutov introduced a spherical lens at the front of the telescope.

Some Maksutov variants actually make the center of the lens partially aluminized, turning the lens into a mirror that reflects the image back. (Think of two-way mirrors, where some light can pass through and some is reflected back, like in your favorite police interrogation scenes.) Such a design can be used to create a *Maksutov-Cassegrain telescope*.

Invisible Astronomy: Looking Beyond the Visible

When most people think of telescopes, they think of people looking through an eyepiece at the night sky. When modern astronomers, astrophysicists, and cosmologists think of telescopes, they're thinking of something very different: seeing the invisible.

Some types of light aren't actually visible. These portions of the electromagnetic spectrum (head to Chapter 3) are abundant in the universe, but humans didn't evolve to see them. Still, we have built telescopes able to detect them.

Toward the beginning of the 1800s, astronomer William Herschel had discovered "invisible light" beyond the red end of the spectrum. This light was what's now known as infrared radiation, but at the time, no one knew what to make of this finding. The electromagnetic theories of James Clerk Maxwell developed this idea throughout the 19th century. In 1887, Heinrich Hertz discovered radio waves. Fast-forward to 1932, when Bell Telephone Laboratories employee Karl Jansky stumbled upon radio interference on his antenna, accidentally creating the first known radio telescope.

Within recent years, telescopes have explored the depths of space at all wavelengths:

- Hubble Space Telescope: Ultraviolet and visible-light spectrum
- Galaxy Evolution Explorer: Ultraviolet
- Compton Gamma-Ray Observatory: Gamma-ray spectrum
- Fermi Gamma-Ray Space Telescope: Gamma-ray spectrum
- Chandra X-ray Observatory: X-ray spectrum
- Spitzer Space Telescope: Infrared spectrum
- Herschel Space Observatory: Infrared spectrum
- Planck Telescope: Microwave spectrum

In the last century, telescopes have transformed into massive data-collection instruments, with humans involved only after the data has been sifted through by computer systems. Many of the most well-known telescopes of our age float in the depths of space and no human ever looks through an eyepiece.

When One Telescope Just Won't Do: The Interferometer

Sometimes a single telescope isn't enough to gather the required data, so astronomers use measurements taken by a variety of different instruments to create a result that contains far higher resolution. A device that's composed of multiple instruments is called an *interferometer*. In astronomical applications, interferometers often consist of a large array of telescopes or antennas set up near each other. (I discuss some non-astronomical interferometers in Chapter 11).

One of the most well known of these facilities is a radio interferometer with the descriptive (but not terribly creative) name of the Very Large Array (VLA). The VLA is run by the National Radio Astronomy Observatory. It consists of 27 radio antennas, each 25 meters in diameter, laid out in a Y-shaped configuration on the Plains of San Agustin, New Mexico. You can find out more about the VLA on its official website (www.vla.nrao.edu/).

Part VI

More Than Just Images: Getting into Advanced Optics

The 5th Wave   By Rich Tennant

Dang kids in the Optics Department, messin' with old Mitch's flashlight again!

OPTICS DEPT RM 201

Part VI
More Than Just
Images: Getting into
Advanced Optics

In this part . . .

This part explores some systems of optics that are significantly more complicated than what you see elsewhere in this book. Most of the systems you work with deal with the index of refraction as a constant for a particular wavelength. This part shows you ways that you can change the index of refraction — an area of study called *nonlinear optics* in which many technological advances are taking place. I present three ways to change the index of refraction, and I describe some situations where these alterations are used. This part also looks at the rather unexpected effects that occur when you have only one or two photons; these phenomena will have a dramatic effect on the way computers process information and the way information is sent in the future.

Chapter 20

Index of Refraction, Part 2: You Can Change It!

. . . .

In This Chapter

▶ Using electric fields, sound, and light to change the index of refraction

▶ Checking out basic devices that use refraction changes to send information

. . . .

The *index of refraction* is typically a quantity that depends on the material and is constant as long as you don't change the wavelength of the light you send through the material. However, you can change a material's index of refraction by altering the material's structure. Using this index change with laser beams, where the light is concentrated in a beam and traveling in more or less the same direction, you can construct devices that can put information on a laser beam by using electric fields, mechanical stress, or even the light itself to shift the structure of the material.

In this chapter, I describe the basic procedures for changing the index of refraction in crystalline materials with electric fields and with mechanical stress through the use of acoustic waves. I introduce some common devices that use these properties for specific purposes, but mostly for *modulating* (putting information on) a laser beam, which means that the irradiance of the light changes in proportion to a signal applied to the material as the light travels through the material. I also discuss how light can change the frequency of light in a material.

Electro-Optics: Manipulating the Index of Refraction with Electric Fields

Electro-optics is the phenomenon where an electric field applied to a material produces a change in the index of refraction of a material. When relatively large electric fields are placed across a material, the structure of the material shifts slightly because it consists of positively and negatively charged particles that are sensitive to electric fields. The positively charged parts move slightly in one direction, and the negatively charged parts (the electrons) move in another direction. All the atoms in a material are bound in some fashion to the atoms around them, so when you shift one, the nearby atoms are affected as well. This shift in the structure produces a change in the index of refraction.

Because the shifts produced by an externally applied electric field (not from the light) in the material are very small, the index of refraction changes by a small amount, on the order of 1×10^{-5}. This change may not seem like it's large enough to do anything, but it's more than enough to produce a noticeable effect on properly conditioned light.

Because shifting one atom affects all the neighboring atoms, the change in the index of refraction can generally happen in any direction, not just in the direction of the electric field.

In this section, I explain dielectric polarization, which is what produces the electro-optic effect. I also describe two types of electro-optic effects and examine some electro-optic devices.

Dielectric polarization: Understanding the source of the electro-optic effect

The electro-optic effect comes from the dielectric polarization of the material. *Dielectric polarization* is, conceptually, a measure of the material's response to an applied electric field. When an electric field is applied to a material, the structure of the material shifts slightly due to the charged nature of the particles that make up an atom. Dielectric polarization measures this shift and its consequence — the change in the material's index of refraction.

You can easily understand dielectric polarization in terms of a *capacitor,* an electrical device used to store electrical charge. A capacitor consists of two conductive plates separated by an insulating material. (Chapter 14 discusses the characteristics of these types of materials.) When you connect a capacitor to a battery, electrons move from one plate to the opposite plate, creating an electric field between the two plates.

In the insulating material, the electrons are bound to a particular position, but the electric field still causes them to move very slightly in the direction opposite to the field. The positively charged parts of the atoms or molecules in the material shift slightly in the direction of the field. This shift creates a slight separation of charge in the material, resulting in a field that reduces the total field between the plates. This charge separation induced by the electric field is called dielectric polarization to distinguish it from optical polarization (which I discuss in Chapter 8).

Dielectric polarization depends on many factors that scientists and engineers study to find optimum performance in particular applications. In general, the dielectric polarization is influenced by many effects. In typical physics fashion, the equation for the dielectric polarization separates out these effects and arranges them in order of what you're most likely to observe. The general equation of dielectric polarization in a general material is

$$P = A_1E + A_2E^2 + A_3E^3 + \dots$$

In this equation,

- P is the dielectric polarization in the material, typically in units of coulombs/meter squared.

- A_i are the *proportionality constants* that decrease in magnitude with higher order. The subscript $i = 1, 2, 3$ and so on represents the numbers of the A constants in the equation. You can often find the proper constant in a table of optical properties of materials.

- E is the electric field in the material.

As a matter of practice, the dielectric polarization causes a change in the index of refraction, so the index of refraction has a power series dependence on the field as well. Most of the time, the absolute value of the index of refraction (the particular number) isn't as important as the *change* in the index of refraction. The following equation lets you determine that change:

$$|\Delta n| \approx C_1E + C_2E^2 + \dots$$

In this equation,

- $P =$ is the absolute value of the change in the index of refraction of the material.

- C_i represents the proportionality constants that decrease in magnitude with increasing order i. You can find these constants in a table of optical properties of materials.

- E is the electric field in the material.

Linear and quadratic: Looking at the types of electro-optic effects

Practical electro-optic devices use the change in the index of refraction produced by an applied electric field to change the properties (usually the optical polarization state or orientation) of a laser beam. Based on the series expansion of the dielectric polarization, you can find two types of electro-optic effects: linear and quadratic. The differences between these categories depend on the symmetry of the crystals involved.

Most electro-optic materials are *crystalline* materials, which means that they have a repeated or *periodic* structure throughout their volume. These materials are relatively easy to analyze and fabricate, so they're what I cover in this section. However, you can also use polymer materials to produce similar effects to crystals, but with the advantage of being able to use electrical fields that have smaller amplitudes. But the polymers typically don't last as long (they break down and no longer function) as the crystalline materials.

The linear electro-optic effect: Pockels effect

The *linear electro-optic* effect, also called the *Pockels effect,* is generally observed in birefringent materials. *Birefringent* materials have two indexes of refraction, determined by the orientation of linear polarization states of the incident light (see Chapter 4 for more about birefringence). However, the Pockels effect doesn't appear in all crystalline materials; only those crystals that are classified as noncentrosymmetric exhibit the linear electro-optic effect. A *centrosymmetric* crystal is one that has *inversion symmetry* — that is, the material looks the same from one direction as it does if you turn it 180 degrees. Some crystals lack this inversion symmetry, and they're the only ones that exhibit the linear electro-optic effect.

The only reason you're interested in this phenomenon is to produce a change in the index of refraction so that you can change the polarization state of incident light in a laser beam. After considering the dielectric polarization in the material and applying some mathematical tricks, the change in the index of refraction for the linear electro-optic effect is

$$|\Delta n| \approx \frac{1}{2}m_0{}^3 E$$

In this equation,

- ✔ $|\Delta n|$ is the change in the index of refraction produced in the material.
- ✔ r is the linear electro-optic or Pockels coefficient, which is a physical characteristic of the material.
- ✔ n_0 is the original value of the index of refraction before the electric field was applied.
- ✔ E is the electric field that is applied to the crystal.

As I discuss in Chapters 8 and 9, the phase retardation between orthogonal components of the optical electric field determines the polarization of the light. Most electro-optic materials are birefringent, so you can analyze linearly polarized light by splitting it into components parallel to each of the principal directions — the fast and the slow axes.

Waveplates (see Chapter 9) are devices that produce a certain amount of phase retardation between the two components. This retardation can produce a change in the polarization state of the light. The electro-optic effect in a material functions as a variable waveplate, in that the amount of phase retardation produced between the components of the light depends on the magnitude of the applied field. Combining the equation in Chapter 9 for phase retardation with the Pockels effect-induced index of refraction change, you get the following:

$$\Gamma = \frac{2\pi}{\lambda_0} r n_0{}^3 EL = \frac{2\pi}{\lambda_0} r n_0{}^3 V$$

In this equation,

- Γ is the phase retardation produced in the material.
- λ_0 is the wavelength of the light in vacuum.
- r is the electro-optic or Pockels coefficient.
- n_0 is the original value of the index of refraction before the electric field was applied.
- E is the electric field that is applied to the crystal.
- L is the length of the crystal that the light travels through.
- V is the electrical potential difference (voltage) applied to the crystal.

This phase-retardation expression is more commonly used than $E = V/L$ because power supplies tell you the potential difference they're providing, not the electric field produced. Calculating $E = V/L$ takes time and is really not that important. The reformulation makes the equation more useful in regards to the signal sent to the electro-optic crystal (which is always given in terms of the electrical potential difference).

The quadratic electro-optic effect: Kerr effect

Crystals that are centrosymmetric can still demonstrate an electro-optic effect. Because they don't have the linear electro-optic, or Pockels, effect, the next term in the index change equation becomes dominant. This effect is called the *quadratic electro-optic* or *Kerr effect,* because it depends on the square of the electric field.

As with the linear effect, you're interested in the change in the index of refraction produced in the material, which is given by the following equation:

$$\Delta n = K \lambda_0 E^2$$

In this equation,

- Δn is the change in the index of refraction produced in the material.
- K is the *Kerr constant,* which is a physical property of the material. You can find this constant in a table of optical material properties.
- λ_0 is the wavelength of the light in vacuum.
- E is the electric field that is applied to the crystal.

Just like with the Pockels effect, the electric field produces a change in the material's index of refraction, which results in a phase retardation between orthogonal polarization states of the light traveling through the material. In a material that exhibits the Kerr effect,

$$\Gamma = \frac{2\pi}{d^2} K V^2 L$$

In this equation,

- Γ is the phase retardation produced in the material.
- d is the distance between the electrodes (the plates that apply the electric field to the crystal).
- K is the Kerr constant.
- V is the electrical potential difference applied to the crystal.
- L is the length of the crystal that the light travels through.

A big difference between the two types of electro-optic effects is their dependence on the orientation of the incident linearly polarized light. The Kerr effect is easier to use because as long as the incident light isn't linearly polarized parallel to the electric field or traveling parallel to the electric field, the device works fine. The disadvantage is that it requires much higher voltages than the linear effect. The linear effect is more complicated to use because the electric field not only changes the index of refraction but can also change the orientation of the principal axes. If linearly polarized light is parallel to one of the principal axes, no phase retardation takes place, and you may be left wondering why your system isn't working.

Examining electro-optic devices

The electro-optic effect is most useful when working with laser beams. As such, the most common electro-optic devices are designed with particular applications of laser beams in mind. In this section, I discuss some of the most common electro-optic devices used with lasers.

The electro-optic modulator: Changing irradiance to write on a laser beam

The most common use of the electro-optic effect is as an optical modulator. An *optical modulator* is a device that can change the irradiance of the light that passes out of the device; it's used to place digital information on a laser beam. (Modulators are one of the components of fiber-optic communication link; see Chapter 15). The modulator works with either of the electro-optic effects covered in the earlier section "Linear and quadratic: Looking at the types of electro-optic effects."

An electro-optic modulator usually has two crossed polarizers, one each at the front and back of the electro-optic material. When you apply no voltage to the crystal, no light passes through the second polarizer. As you increase the voltage, more light passes through the second polarizer, reaching a maximum and then going back to zero as the voltage increases. Because the electro-optic material functions as a variable waveplate, the polarization state or orientation is constantly changed, causing the change in the amount of transmitted light. This process is one way to produce an optical signal where the maximum irradiance corresponds to a one and no light corresponds to a zero.

Because the light is an electromagnetic wave and the phase retardation is periodic, the transmitted irradiance varies in a periodic fashion based on the magnitude of the potential difference applied to the material. For an electro-optic modulator, the transmitted irradiance is given by

$$I = I_{\max} \sin^2\left(\frac{\pi}{2} \frac{V}{V_\pi} \right)$$

In this equation,

- I is the transmitted irradiance coming out of the modulator.

- $I_{\max}$ is the maximum possible transmitted irradiance.

- V is the electrical potential difference (voltage) applied to the material.

- V_π is the voltage necessary to produce a phase retardation of π.

The electrical potential difference is generally used because that's the quantity measured by power supplies or volt meters. It's an easier quantity to deal with in general, so the transmitted irradiance is given in regard to the potential difference instead of the electric field magnitude. The quantity V_π, also called the *half-wave voltage,* is an important characteristic of any electro-optic device. Its particular value depends on electro-optic constants of the material, the direction of the field applied to the material, and the direction of propagation of the light through the material, so it's very handy to know. V_π is the amount of voltage necessary to produce a phase retardation of π, or $\lambda/2$, between the two orthogonal components of the optical electrical field. With this voltage, the electro-optic material functions as a half-wave plate (hence the term half-wave voltage).

Using an electro-optic modulator in a fiber-optic communication link requires that the electrical signal correspond exactly to the optical signal placed on the laser beam. In order to ensure this linear relationship, you place a quarter-wave plate between the electro-optic material and the first polarizer. This plate causes the transmittance to be 50 percent of the incident light with no voltage applied, but it places the modulator in the linear part of the sine-squared curve. This way, a small voltage applied to the electro-optic material results in an exact replica on the laser beam — the optical signal. Without this bias, the optical signal picks up many nonlinear components, which function basically as junk. You don't want to talk to your grandmother over such a fiber-optic link — the connection sounds like you're talking with an alien.

Liquid crystal displays

Liquid crystal displays (LCDs) are another type of electro-optic device, but they don't work the same way as the other electro-optic effects in crystals. *Liquid crystals* are long, rigid molecules that are free to change their orientation randomly. When you apply an electric field to them, they tend to line up with the field. This alignment changes the optical properties (particularly the index of refraction) of the liquid, so this electric field application can change the polarization state or the orientation of the light. When you remove the electric field, the crystals return to a random orientation.

Many calculator displays utilize this process. The LCD contains a liquid crystal material sandwiched between a reflector on the bottom and a linearly polarizing film on top. When no power is coming to the screen, the display appears gray or green depending on the type of reflection material used at the bottom. Turning on the calculator applies an electric field to a particular region in the display, and the liquid crystals in that region orient in a uniform fashion around the electric field and change the polarization of the light reflected off the bottom. When this light hits the top of the screen, it's blocked by the polarizing film. This blockage is why the numbers or letters on the screen look black. When the electric field is removed, the light is no longer rotated and passes through the last polarizer.

LCD televisions use a similar but much more complicated procedure. You don't simply turn the light on and off; you have to worry about color, so the LCD TV screen controls pixels that allow different colors (usually red, blue, and green) through. Image information is used to drive the liquid crystal elements to produce the pictures that you see.

Acousto-Optics: Changing a Crystal's Density with Sound

Mechanical pressure can change the index of refraction of a material. You can think about this process like squeezing a crystal between your fingers. Although this experiment does generate a change in the index of refraction, actually building a stable optical system around someone holding a crystal is difficult, even if you use a C-clamp rather than your fingers; you can only loosen or tighten the screw so fast.

Fortunately, you have another option, another phenomenon that involves pressure: *acoustic waves,* or sound waves. Sound is a mechanical wave that propagates through materials, such as air, by periodic changes in the air's density. These decompressed and compacted regions of the air travel from the source of the sound to your ear, where your brain interprets the sound. Acoustic waves can exist in any material, so you can apply them to a crystal.

The acousto-optic effect: Making a variable diffraction grating

Acousto-optic devices use crystals but don't involve the dielectric polarization characteristics that electro-optic devices do (see the earlier section "Electro-Optics: Manipulating the Index of Refraction with Electric Fields"). Acousto-optic devices basically work by using a crystal fitted with a transducer to periodically squeeze and release the material. A *transducer* is a device that vibrates to produce sound from an electrical signal input; the frequencies are typically in the megahertz range, so you don't hear anything, but light can detect them.

The acoustic waves in the crystal periodically change the density of the material and the index of refraction because those characteristics are nominally proportional. With a periodic variation in the index of refraction, the material behaves like a diffraction grating (see Chapter 12). Even though the sound waves are traveling in the material, light travels so much faster than sound that the light sees a stationary grating, for most practical purposes.

The diffraction grating equation I present in Chapter 12 holds for the acousto-optic effect. The light hits the acoustic grating and diffracts into various orders. The *zeroth order,* the one that travels straight through, is modulated, but it appears as a small ripple on top of a large, constant signal. The first diffracted orders can also work, but their irradiance is lower and is affected by the amplitude of the acoustic wave. The larger the acoustic signal (the harder you drive the transducer), the greater the amount of energy sent to the diffracted orders, and vice versa.

Using acousto-optic devices

Because the acousto-optic effect creates a variable diffraction grating in the material, you can use it for a variety of schemes. A couple of common uses are beam scanners and optical modulators.

Acousto-optic scanner

The *acousto-optic scanner* is much more robust scanner than a mechanical one simply due to the fact that the former has no moving parts. You can find acousto-optic scanners in bar-code scanners (like in a grocery store) or in various quality control arrangements in industry.

Acousto-optic scanners utilize diffraction gratings (see the preceding section) to deflect a laser beam, and varying the acoustic frequency changes the deflection angle. According to the diffraction grating equation in Chapter 12, the diffraction angle of any particular order depends on the *grating period,* or grating spacing. Because the transducer is responsible for making the grating, the frequency of the electrical signal sent to the transducer is variable. For the scanner, the wavelength changes inversely with the frequency change of the acoustic wave. Therefore, the grating spacing (the acoustic wavelength) also adjusts, so the position of a particular diffracted order of the transmitted light shifts as well. If the frequency of the electrical signal is varied constantly over a certain range, the diffracted beam sweeps through a particular angle, making a beam scanner.

Acousto-optic modulator

You can use the acousto-optic effect as a versatile optical modulator. Because of the flexibility that comes with the transducer and the resulting variable diffraction grating, the acousto-optic effect can put a signal on a laser beam (called *modulating* the beam) by varying the amplitude or the direction of the beam. The irradiance of the light in a particular diffracted order depends on the size of the index of refraction change, which depends on the transducer amplitude. The larger the amplitude of the acoustic wave, the greater the irradiance of the diffracted order. A detector monitoring the diffracted beam receives the optical signal carried in the irradiance of the light incident on the detector. This modulation is very much like the signals used for AM radio.

Shifting wave frequency with the Doppler effect

A subtle effect that occurs with light bouncing off traveling acoustic waves is the *Doppler shift* of the light. Basically, if light bounces off a mirror that's moving toward it, it gains energy from the bounce, and its frequency shifts slightly higher. If the light bounces off a mirror that's moving away from it, it loses energy from the bounce, and its frequency shifts slightly lower.

A similar effect occurs in an acousto-optic device, but the light bounces off an index change

(the acoustic wavefront), not moving mirrors. Therefore, the diffracted orders the device produces also shift slightly in frequency. The shift is small but detectable, so acousto-optic modulators can also frequency modulate an optical beam, such as the radio signals for an FM radio station. Additionally, acousto-optic modulators find many uses in laser systems because of this Doppler effect, especially in the area of creating pulsed laser systems.

Another way to modulate a laser beam is to vary the location of a particular diffracted order by varying the frequency of the signal sent to the transducer. Changing the frequency changes the grating spacing and the diffraction angle. This method is basically like jiggling the laser beam around a detector in such a way that you send information.

Frequency Conversion: Affecting Light Frequency with Light

Dielectric polarization (which I discuss earlier in the chapter) not only works for electric fields that you can apply to a material but also describes the effect of any electric field, including the incident light's electric field. One of the biggest innovations for laser systems involves the ability of light and a material to interact in a way that produces light of a different frequency. This effect comes from the dielectric polarization, with the electric field supplied by the light itself. This phenomenon can create a large number of outcomes, and the following sections introduce some of the most common.

Second harmonic generation: Doubling the frequency

The first frequency conversion process observed is called *second harmonic generation* (SHG). In this process, long-wavelength light, usually in the near-infrared part of the electromagnetic spectrum, is sent with a rather large irradiance into a special crystal. The high irradiance is necessary to drive the material hard, causing the electrons to move significantly.

Because the atoms are fixed in place (because the crystal is a solid), this intense delivery causes them to oscillate or wiggle around their equilibrium positions *asymmetrically* (they may be able to move a greater distance in one direction than another). This type of motion is called *anharmonic motion*.

You can think of anharmonic motion as what happens if you take a spring and stretch it so that it's deformed (you have exceeded its elastic limit). If you hang a weight from the end of that spring and let it move up and down, you see that it moves a much smaller distance in the upward direction than it does downward. Anharmonic oscillations look similar. When this oscillation happens inside a crystal, some new light is observed at the output side of the crystal. A lot of the light has the same frequency as the incident infrared light, but some of the light has twice the frequency of the incident light (hence the term second harmonic generation).

SHG plays an important role in many solid-state laser systems today. As I mention in Chapter 18, the most efficient high-powered lasers typically emit light with a wavelength in the 1-micrometer (infrared) range, but some applications, especially medical ones, require blue or green light. Lasers in the blue and green wavelength ranges aren't very efficient, so very few appropriate systems are available. However, with the advancement in SHG technology (primarily in the area of material design and fabrication), special SHG crystals can be placed inside the laser cavity and produce the necessary blue or green light.

Parametric amplification: Converting a pump beam into a signal beam

In some laser applications, the necessary wavelength isn't available with second harmonic generation. Another frequency-conversion process, called *parametric amplification,* involves a laser beam with a large amount of power (called the *pump beam*) and a very weak laser beam (called the *signal beam*) with the wavelength that you need. In this process, you transfer light from the pump beam to the signal beam by converting some of the pump beam into the signal beam (not by simply taking light from the pump and adding to the signal beam). In other words, you convert light of the pump-beam frequency to the signal-beam frequency, increasing the power in the signal beam.

In order for parametric amplification to work, a third beam must be present. This beam, known as the *idler beam,* is actually produced by the difference in frequency between the pump and signal beams. So in an actual parametric amplification process, you end up with the three different frequencies of light: the pump frequency, the signal frequency, and the idler frequency. Parametric amplification is actually very useful for modern tunable laser systems because the idler provides other frequencies that other frequency conversion processes can use.

Sum and difference frequency mixing: Creating long or short wavelengths

If you have two different lasers that have rather large intensities, you can use them in a frequency-conversion process called *frequency mixing* in appropriate crystals. If you have two different frequencies, the oscillations set up in the crystal can work to produce light that has a frequency equal to the sum of the two frequencies or a frequency equal to the difference between the two. (SHG is an example of sum frequency mixing where the incident beam is treated as two beams with the same frequency.) This process is very handy if you need really long-wavelength light *(difference frequency mixing)* or rather short-wavelength light *(sum frequency mixing)*. Which process you get is determined by the wavelengths involved and the properties of the crystal used. Combined with parametric amplification, these processes have revolutionized the applications of lasers by making continuously-tunable frequency lasers.

Phase matching

Although my presentation of frequency conversion seems rather simple, frequency conversion isn't just a matter of sending laser light into a crystal; you must put much careful thought into creating the situation where any of these processes can occur. The biggest consideration is a concept called *phase matching,* which helps combat the destructive interference that compromises your conversion. To get a frequency conversion process to happen in a crystal, you need to find wavelengths that have the same index of refraction (that is, light beams with the two different wavelengths travel with the same phase velocity along a particular direction in the crystal). Without phase matching, destructive interference significantly reduces (or eliminates) the amount of the frequency-converted light.

You can produce phase matching in a variety of ways, but the simplest method involves beams that are linearly polarized and orthogonal to each other. In a birefringent material, the ordinary polarized light at one frequency can have the same index of refraction as the extraordinary polarized light (see Chapter 4 for a discussion about ordinary and extraordinary polarized light). The index of refraction can't be the same for two different frequencies if both beams have the same polarization.

Chapter 21

Quantum Optics: Finding the Photon

....

In This Chapter
▶ Examining experiments supporting the dual nature of light and matter

▶ Exploring applications of quantum entanglement

....

*M*ost of this book deals with large amounts of light (whether waves, particles, rays or beams) and how you can make the light do what you want — everything from forming images to cutting metal. This chapter gives you a glimpse of some of the more unexpected properties of light and what you can use those properties for.

In this chapter, I describe the dual nature of light and matter and describe the results of experiments, based on work with electrons, that show that photons have particle and wave properties. Just like their matter counterparts, photons don't behave like you may expect them to when you get down to one or two photons. I end with a bizarre occurrence where photons can seemingly communicate with each other instantly.

Weaving Together Wave and Particle Properties

Albert Einstein proposed that light is localized, not a wave of infinite extent. It still has wave properties, but it's particle-like in that the waves of light are finite in length, resulting in their energy always being absorbed in certain amounts at particular locations. He called this idea *quanta;* many contemporaries called it a *corpuscle,* and later the name was replaced with its current label: *photon.*

So classifying light as a *wavelet* (a wave with a definite length) means that the light not only has wave properties (can exhibit interference with other photons and undergo diffraction) but also has particle-like properties (has energy that can be deposited all at once, as demonstrated in the photoelectric effect). Light isn't quite so unusual, because particles of matter have also been shown to have both particle- and wavelike properties; the different properties are just easier to see for light.

The following sections summarize the wave and particle properties by showing you the similarities between light and matter. These findings, based on two men's hypotheses, started out seeming to be quite unlikely, but numerous experiments have proven them correct.

Seeing wave and particle properties of light

The results of the photoelectric effect (see Chapter 3) were giving scientists problems. In a famous experiment, scientists shone light on a metal plate; when the plate's electrons absorbed enough energy from the incident light, the electrons were emitted from the plate. Surprisingly, researchers found that changing the light's irradiance didn't affect the amount of the electrons' kinetic energy (the speed at which the electrons left the plate), and electrons that were emitted left the plate immediately; energy didn't build up over time. The frequency of the incident light, not its irradiance, determined the electrons' energy. The wave theory of light, the go-to explanation at the time, couldn't explain these results.

In 1905, Albert Einstein suggested that light waves aren't like classical waves. The classical picture of light waves was that the waves were infinite in extent; a light wave constantly deposits energy until something turns the wave off. Einstein proposed that light is indeed a wave, but not infinite in extent, so that the light quanta deposits their energy all at once. The present idea is that the quanta are rather like *wavelets* (small waves where small means finite length, not small amplitude; see Chapter 3 for more about this idea).

To account for the observations made in the photoelectric effect, Einstein proposed two ideas for photons. The first was that even though photons don't have a mass, they still have energy that is contained in the wavelet. This energy is given by

$$E = hf = \frac{hc}{\lambda}$$

In this equation,

- E is the energy carried by the photon.
- h is Planck's constant and is equal to 6.626×10^{-34} joule-seconds.
- f is the frequency of the photon.
- c is the speed that light travels in a vacuum and is equal to 3.0×10^8 meters per second.
- λ is the wavelength of the photon.

The second idea was that, despite having no mass, photons must have momentum because they carry energy. The following formula calculates this momentum:

$$p = \frac{E}{c} = \frac{h}{\lambda}$$

In this equation,

- p is the momentum carried by the photon.
- E is the energy carried by the photon.
- c is the speed that light travels in a vacuum.
- h is Planck's constant.
- λ is the wavelength of the photon.

Although these two ideas explained all the problematic observations in the photoelectric effect, they also showed that the wave and particle theories weren't mutually exclusive. For many years, scientists thought that light must be one or the other because daily experience indicated that things in the mechanical world are either waves (like sound waves, water waves, or waves on strings) or particles (like tiny marbles). Nothing to that point showed the property of being a wave in one case and a particle in the other. Although experience suggests that these two properties can't exist at the same time for any object, Einstein's ideas showed that photons simultaneously have wave properties (frequency and wavelength) and particle properties (energy and momentum).

In your day-to-day experiences, you don't notice the graininess of the light around you because the number of photons present is so large that they appear to be like continuous waves. Similar to quantum mechanics, you notice the particle properties only when you can look at a small number of particles individually rather than the average effects of a large number of photons.

Looking at wave and particle properties of matter

In 1923, not long after Einstein proposed his quantized light waves (see the preceding section), Louis de Broglie postulated that because photons simultaneously have wave and particle properties, matter likewise simultaneously has wave and particle properties. Using Einstein's idea about momentum and energy for photons, de Broglie suggested similar properties for particles such as electrons, protons, neutrons, atoms, and so on.

In particular, de Broglie proposed that because the particle properties (namely energy and momentum) are well defined, the wavelength of a particle is

$$\lambda = \frac{h}{p} = \frac{h}{mv}$$

In this equation,

- ✔ λ is the wavelength of the particle of matter, called the *de Broglie wavelength*.
- ✔ h is Planck's constant and is equal to 6.626×10^{-34} joule-seconds.
- ✔ p is the momentum carried by the particle.
- ✔ m is the mass of the particle.
- ✔ v is the speed of the particle.

In a similar fashion, de Broglie suggested that the frequency (the other wave property) of the particle is

$$f = \frac{E}{h}$$

In this equation,

- ✔ f is the frequency of the matter wave (particle).
- ✔ E is the kinetic energy carried by the particle.
- ✔ h is Planck's constant.

Seeing with electrons: The scanning electron microscope

You may have heard about scanning electron microscopes (SEM) and seen pictures taken of dust mites or housefly eyes that show much smaller details than a typical microscope can. A SEM takes advantage of the shorter wavelength of electrons to enable you to see such small details.

According to Louis de Broglie's formula (see the nearby section "Looking at wave and particle properties of matter"), the faster electrons move, the shorter their wavelength. In a SEM, electrons are accelerated through a potential difference of many thousands of volts, giving them a wavelength about 100 to 1,000 times smaller than that of visible light. Instead of using glass lenses (which would absorb the electrons or scatter them away), scientists use magnetic fields to concentrate the electrons onto the surface of an object. The electrons bounce off the surface and are collected by a detector. The electron beam is scanned across the object to build up an image, similar to what is presented with light in Chapter 7. Because you can't see electrons, you use another type of detector to record the position of the electrons; a computer reads the data from the detector and presents you with an image you can see.

de Broglie's relationships showed that matter particles have particle properties (energy and momentum) and wave properties (frequency and wavelength) simultaneously and that the two properties are complementary, not mutually exclusive as everyone had thought previously.

As with light, you don't see the wave properties of matter under everyday circumstances. You have to have very special situations, like very small apertures or slits, to be able to detect the wave properties. When trying to look at the quantum properties of matter or light, making single particles, such as electrons, is easier than making single photons; however, seeing the wave properties of matter is usually much more difficult than doing the same for light because of the extremely small scale required.

With light, you know what is waving: the electric and magnetic fields. With particles, however, exactly what's waving it isn't clear. In quantum mechanics, the wave function for a particle is a probability distribution function. In other words, what is waving is the likelihood of finding the particle in a particular location. This aspect of matter waves is still rather baffling to many people, including me.

Experimental Evidence: Observing the Dual Nature of Light and Matter

To see the particle properties of light, you need only one photon at a time, but isolating only one isn't easy to do. However, you can more easily get one particle (say, an electron) at a time and do experiments with it. Because both matter and light have dual natures, experiments with single particles show what would happen in similar experiments with single photons. The following sections introduce some experiments that show how experiments that work with properties of light can help illuminate (pun intended) properties of matter.

Young's two-slit experiment, revisited

In Chapter 11, I talk about what happens when light passes through two closely spaced slits. A modulated irradiance pattern — bright and dark fringes that are equally spaced — appears on a screen opposite to the slits. Numerous experiments have shown that electrons sent through two closely spaced slits end up producing a similar pattern at a detector (your eyes don't detect electrons like they do light, so you can't use a screen).

In the electron version of the two-slit experiment, an oven emits electrons, and then a velocity selector chooses electrons that have a particular speed (which determines the energy of the electrons so that you can calculate the frequency and wavelength). The electrons then pass through the two closely spaced slits. A small electron detector is moved in a line parallel to the plane that contains the slits and detects the electrons at particular points along this line.

If you send a small number of electrons per unit time through the slits and let the detector just count the number of electrons detected over a certain period of time, you will find that at some points, nearly no electrons are detected and at other points, a relatively large number of electrons are detected. This result is the equivalent of the interference pattern produced by light, just with much smaller fringe spacing. As in the light experiment in Chapter 11, you can find the irradiance pattern that the electron waves produce; you use the interference of two electron wave functions to predict the locations where the electrons will go.

Now, if you reduce the number of electrons to one at a time, an interesting effect happens: Placing a detector at a location that corresponds to a point where almost no electrons are detected (equivalent to a dark fringe when you use light), you very rarely detect an electron, even if you leave the detector in that position for thousands of individual shots. If you move the detector to a location that corresponds to a point where a large number of electrons are detected (equivalent to a bright fringe when using light), you get a very large count when you leave the detector in that position for a large number of individual shots. Because you're shooting one electron at a time through the slits, the only explanation for the fringe pattern is that each electron went through both slits! How can that be possible? It can only happen if the particles have wave properties, because interference (the appearance of count fringes) only happens with waves. So this wave property of electrons can do something else remarkable; matter waves can sense both slits and make the electron behave accordingly.

Currently, scientists have the ability to generate and detect a single photon. If you repeat the two-slit experiment with light by sending just one photon at a time, you end up with the same result that happens with the electrons, including the very small/very large photon counts in the fringe areas and the mysterious trajectory through both slits.

Diffraction of light and matter

One of the first experiments to prove de Broglie's hypothesis about the wave properties of matter involved looking at what happens to electrons that are sent into crystals, which have an arrangement of atoms that repeats throughout the volume. (See the earlier section "Looking at wave and particle properties of matter" for more on de Broglie's ideas.) The phenomenon is called *electron scattering* because the electrons don't bounce off a surface but rather off atoms (see Chapter 4 for more details about scattering). C.J. Davisson and L.H. Germer performed numerous experiments with electrons and crystals and noticed that increased electron counts occurred at particular angles, a result that could only be explained by the electrons hitting a diffraction grating. The closely spaced planes of atoms in the crystals functions like diffraction grating does with light. Analyzing the angle and the spacing of the atoms that did the scattering led to the first experimental confirmation of the wavelength of electrons, which exactly matched the predictions of de Broglie's hypothesis. Only the idea that electrons have wave properties can explain the appearance of concentrated counts in some areas and virtually none elsewhere.

X-rays also showed a similar bright-spot pattern as the electron experiments did. Scientists could determine the spacing of the atoms in the crystal materials based on the wavelength of the x-rays and the angles where the bright spots appeared. This process is called *x-ray diffraction* and is an important tool in the area of materials research, which provides scientists with information about the structure of the materials.

The Mach-Zehnder interferometer

As I note in Chapter 11, the *Mach-Zehnder interferometer* is a device that separates the paths of light waves so that the light doesn't travel along the same path twice. You can also set up this interferometer to work with electrons. A beam splitter sends electrons around two different paths, and then the electrons come back together. Although the distances are extremely small (much, much smaller than for visible light), you can obtain interference patterns with the electrons; as with the two-slit experiment (see "Young's two-slit experiment, revisited" earlier in the chapter), the detector still detects more electrons in the "bright fringe" areas than in the "dark fringe" areas. The same experiment also works with photons.

Quantum Entanglement: Looking at Linked Photons

With the single-photon experiments yielding interference patterns that I discuss earlier in the chapter, you don't actually know where the photon is or what path it takes unless you make a measurement. After you make the measurement, you can no longer see the interference because you have determined exactly where the photon is. So until you measure, the photon goes through both slits at the same time because you don't know anything different. This conclusion may seem a bit nonsensical, but that's how typical people studying quantum effects account for it.

However, for a photon to go through both slits at the same time seems to indicate that there's more to this wave property of light. If a single photon can be in two places at the same time, perhaps photons can communicate (for lack of a better word) with each other even though they're separated by a vacuum. You can look at this interaction as a coordination of properties (in particular, polarization) of two or more photons that don't follow random statistics. This property where photons can be linked (somehow) is referred to as *entanglement.*

Spooky action: Observing interacting photons

Interacting photons, or *entangled photons,* are basically two or more photons with a special link. The significance of this link is that if you change a property of one of the photons, the other entangled photons instantly change their state to match. Why entanglement occurs isn't clear, but numerous experimental situations have seemed to indicate that it does happen; it probably has something to do with the interference effects of single photons presented earlier in this chapter. This phenomenon is so mysterious that Albert Einstein called it the "spooky action." Entanglement is one of the subjects in quantum optics that may have a significant impact on scientists' understanding of how the universe works.

Anton Zeilinger, a major researcher in the area of entanglement, likes to use the idea of a pair of dice to illustrate entanglement. In this analogy, the dice represent the photons, and the numbers on the dice represent some property of the photons, such as polarization state or orientation. Zeilinger goes on to say that if you roll the dice and the numbers that come up always match (like a pair of threes or a pair of fours), the dice are behaving as if they were entangled. (Each die is independent and not weighted such that the same number is designed to always come up.) This correlation continues regardless of whether the dice are next to each other or a mile apart.

Recently, many experiments have shown that entangled photons can exchange information instantly even if the distance between them is quite large (even many miles). Numerous processes, including *nonlinear down-conversion* (a process where the wavelength of the incident light on a crystal is converted to a longer wavelength) in crystals such as beta barium borate or potassium titanyl phosphate, can produce entangled photons, but these methods require intense laser beams and special nonlinear crystals. Scientists have recently been able to produce numerous entangled photons by using simple semiconductor devices made from gallium arsenide or indium arsenide. The semiconductor devices are much more attractive because they can make numerous entangled photons almost on demand, which is essential in order to use quantum entanglement in communication or computing (see the following section).

Encryption and computers: Developing technology with linked photons

Although rather bizarre, photon entanglement provides some interesting possibilities for applications, especially in the realm of secure communication and computing.

Super encryption

Using the idea of entangled photons, super-secure communication (surprisingly, even over very large distances) is possible. The release of information depends on who detects the photon:

- ✔ If someone other than the intended receiver intercepts the photons, the data is destroyed and the sender and receiver know that someone attempted to intercept the data.

- ✔ When the intended receiver detects the photon, the data is released because both sender and receiver have the entangled photons and are set to receive the agreed-upon property change, such as polarization state, to transfer information.

This capability is a consequence of the photons, in general, being in multiple states at the same time. You don't know which state a photon is in until you make a measurement, but then you know exactly where it is and what state it's in. Entanglement is the ultimate in secure communication.

Quantum computing

You can also use the multiple-state property of entangled photons for *quantum computing*, which utilizes the indeterminate state of photons (that is, the idea that you don't know what state photons are in until you measure). As with optical computing in general, quantum computing is massively parallel because it doesn't compute just one solution at a time but rather all possible solutions with a given set of inputs.

This computing works even more quickly if you use entangled photons. Right now, the limit to quantum computing is being able to generate enough entangled photons to make the practice useful. Quantum computers may not appear in stores tomorrow, but they are on the way.

Part VII
The Part of Tens

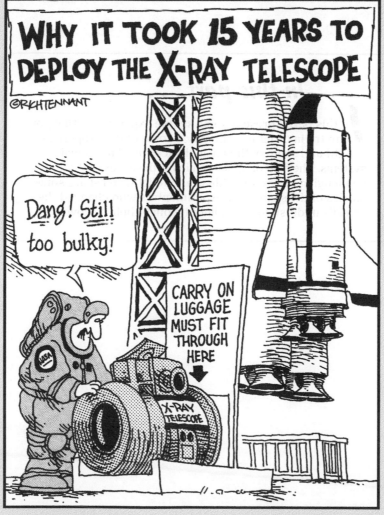

The 5th Wave   By Rich Tennant

In this part . . .

In this part, I present some simple experiments that you can try in order to experience some of the properties of light and some of the basic devices used in optical systems. The field of optics is filled with stories of students challenging the existing ideas about light, so I share some stories of the significant experimenters who changed the world's understanding of light through their discoveries.

Chapter 22

Ten Experiments You Can Do Without a $1-Million Optics Lab

In This Chapter

▶ Using everyday items to demonstrate some optical principles

▶ Seeing optics in common situations

*I*n this chapter, I show you how you can see some of the basic principles of optics at home or in the world around you. Of course, you can find a lab with all the equipment you need to duplicate these tests, but why go to the trouble when examples of optical fundamentals are everywhere in the world around you?

Chromatic Dispersion with Water Spray

You can commonly see *chromatic dispersion,* the separation of light into the separate wavelengths, in the effects of a rainbow or by shining a light through a prism. If you don't have a prism handy, grab a water-filled spray bottle and two good light sources, such as a laser pointer and a flashlight that lets you focus the beam.

Shine one of the light sources across a darkened room. Spray the water along the beam of light to expose as much of the beam as possible. Repeat this process with your other light source and compare the different colors you see based on each light source.

Be careful not to shine the light directly at other people. Any concentrated light can damage the sight of you or other people in the room.

The Simple Magnifier

A water droplet can act as a simple magnifying lens. Water droplets tend to form as spherical droplets under the influence of surface tension. When the drop is attached to an object such as a plant leaf, the drop's spherical shape is distorted but still capable of forming an image of whatever is behind it. You can use this principle to make your own homemade magnifying glass.

Place the surface you want to magnify (such as a piece of paper with a picture or some text) flat on a table or another sturdy surface; wobbling can dislodge your magnifier.

Lay a piece of flat, clear glass on top of whatever you want to magnify. (Be sure that it's positioned over a place that you want to magnify; during the experiment, you can move it around, but picking it up is difficult.) Use an eyedropper or a straw to drop several drops of water onto the glass one at a time. The drops stick together and form a large droplet. Make sure your drop doesn't get so big that it overflows to the side. As you look through the water droplet, everything underneath is magnified to several times its actual size.

Microscope with a Marble

Microscopes, like telescopes or magnifying glasses, use lenses to enlarge objects. Technically, a *magnifying glass* is a piece of glass whose shape lends itself to magnifying images, so surprisingly (or perhaps not), a clear glass marble provides a common, low-grade convex lens that you can use as a magnifier.

The round shape of the marble refracts light (see Chapter 7) and causes objects seen through it to appear larger. Although marbles aren't perfect lenses and can't create scientific-grade microscopes, you can use them to create basic hand lenses for basic study or play.

Put a good clear marble on one corner of a piece of construction paper. Roll the paper into a tight tube around the marble and tape the paper securely so that it doesn't come unrolled. Position the marble so that half of it protrudes from the tube and then look through the open end of the tube to use your microscope.

Focal Length of a Positive Lens with a Magnifying Glass

A magnifying glass is essentially a single element lens with a *positive* focal length, meaning it focuses on a point on the other side of the lens. (Check out

Chapter 5 for more on focal length.) You can easily measure the focal length of a positive focal lens by using a strong light source, a magnifying glass, a piece of paper, and a ruler. You want the light source to be a fair distance away; ideally, it should be as far away as possible, but a light across the room works. *Warning:* For this experiment, don't use the sun as your light source; you'll set the paper on fire.

Project the light from the light source through the lens and onto the paper. Adjust the distance between the lens and the paper until the image is at its sharpest. Now measure the distance from the lens to the image on the paper (have somebody help if necessary). This distance is the focal length of the lens.

Telescope with Magnifying Glasses

You can build a simple telescope with a magnifying glass and a pair of weak reading glasses from a drugstore. The magnifying glass serves as the *eyepiece* (short focal length, high power lens) to your telescope; a single lens from the glasses is the *objective* (short lens). This lens transmits the light from stars or other types of objects that emit or radiate light. Flip to Chapter 13 for more on eyepieces and objectives in lens systems.

You also need to buy or make two rigid tubes, such as cardboard tube or plastic pipe, that can slide one inside the other. The diameters of these tubes must be larger than the largest lens. Remove one lens from the reading glasses and determine its focal point by using the process in the preceding section. Repeat the process for the glass from the magnifying glass.

Add the focal lengths of the objective and the eyepiece. Divide the sum in half and add an inch to find the length of each tube. Cut the two tubes to length; they should be the same size. Cut two equal circles from poster board to serve as mounts for the lenses. The circles must be larger than either lens and fit the ends of the tubes. Cut a hole slightly smaller than the objective in the middle of one circle and glue the edges of the lens to the mount. Repeat this step with the other circle for the eyepiece.

Glue the mount with the eyepiece to the end of the narrower tube and the mount with the objective to the end of the wider tube. Insert the smaller tube into the larger tube. Point your telescope at an object in the distance and look through the eyepiece. Slide the tubes in and out of each other until the object comes into focus. *Warning:* Don't look directly at the sun.

Thin Film Interference by Blowing Bubbles

Thin films and coatings are subject to interference effects (see Chapter 11). Remember that light is reflected at the interfaces of different materials. In the case of a bubble, the reflected light wave from the air/soap interface on the outside of the bubble is out of phase with the reflected light wave from the inside of the bubble/air interface.

To observe interference effects, find a good bubble mix (or make your own) and blow some bubbles; try to leave them on the wand or blow them into a dish of water or soap where they can last longer. As light impinges on the bubble (film), three things happen: A portion of the light reflects off the outer surface; a portion enters the film and then reemerges after reflecting off the second surface; and a portion enters the film, reflects back and forth a multiple odd number of times within the film, and then reemerges. All these reflections interfere and determine what reflected light you actually see.

As each wave traverses the film, it undergoes a phase shift proportional to the thickness of the film and inversely proportional to its own wavelength. At a given film thickness, interference is *constructive* (additive) for some wavelengths and *destructive* (subtractive) for others. White light impinging on the film reflects with a hue that changes with thickness.

As the bubble is thinning due to evaporation, you can observe a change in color. Thicker film cancels out red (longer) wavelengths, causing a blue-green reflection. Thinner walls cancel out yellow (leaving blue light), then green (leaving magenta), and then blue (leaving a golden-yellow), respectively.

Polarized Sunglasses and the Sky

As light reflects off surfaces, a portion of it also scatters. A scattered light wave's electric field is perpendicular to its direction of travel, resulting in a "side to side" direction. This side-to-side polarization of light is what makes polarized sunglasses interesting.

As light scatters off a horizontal surface, the waves' electric field mostly vibrates horizontally. Because large horizontal surfaces (such as bodies of water, roads, and cars) are common outdoors, most outdoor glare comes from these horizontally vibrating electric fields.

Polarized lenses are designed to allow the electric charge of the lens molecules to move more easily in the horizontal direction than the vertical direction. (Chapter 8 gives you the lowdown on polarization.) As the electric fields from the scattered light excite the lens's electrical charges, the lens absorbs energy, allowing fewer horizontally oriented scattered light waves to pass through. With the reduction in scattered light, the eye is able to more clearly process the reflected light.

Look at a patch of clear sky through one polarized sunglass lens while you rotate the glasses. As the lens rotates, some parts of the sky seem to change from brighter to dimmer and back. If you position the lens so the sun is to the left or right of the lens, the sky appears brightest and darkest when the sun is above or below the lens, respectively. The sky has intermediate brightness if you view it through the lenses at any angle in between. Additionally, different regions of the sky look bright or dark as you turn the polarizing lenses in different directions.

Mirages on a Clear Day

Have you ever been driving down the highway on a hot day and seen what appears to be wet pavement? However, you never reach that wet spot because what you see is actually a true mirage commonly known today as a *highway mirage*.

To reproduce the effect, you need a hot plate (or a metal plate on a stove), a laser pointer, some foil, a piece of paper, and a pen. ***Remember:*** You're dealing with hot materials, so use some hot pads and plenty of caution.

Set the laser pointer to shine across the top of the hot plate as close as possible to the surface. Place the piece of paper on the wall where the laser is shining and mark the spot where it hits. Create a foil tent 3 to 4 inches high across the width of the hot plate so that the laser shines through it. Turn on the hot plate and watch the laser dot move on the paper; mark the spot it stops at. Using the hot pads to protect your hands, remove the foil tent and watch the spot move again.

Spherical Aberration with a Magnifying Glass

Primarily, *spherical aberration* (see Chapter 7) is the blurring of an image you're seeing through a lens. Light passing through the center of a lens focuses at the intended point, but light passing through the outer edges of the lens doesn't focus at the intended point because the edges focus at different distances than the center does (a function of lens thickness and shape). Spherical aberration results in a less-clear, less-sharp image.

To easily witness (and fix) spherical aberration, grab your magnifying glass and do a little experimenting. Look through the magnifier at an object and notice the blurriness. Using dark, thick paper, cut out a donut slightly larger than the outside diameter of the magnifying glass; the donut's inside hole should leave at least three-quarters of the glass visible. Place this donut over the lens and look back at the same object to see how this addition affects the clarity and sharpness of the image. You can make a second donut that covers the outside half of the lens and recheck the image. As you eliminate the light passing through the outer edges of the glass, you're eliminating the unfocused light, and the object should become clearer.

Chromatic Aberration with a Magnifying Glass

If you look through a cheap simple magnifying glass, you see a slight rainbow fringe around objects that is the result of chromatic aberration. *Chromatic aberration* (head to Chapter 7) occurs because the index of refraction of all optical materials changes for the different wavelengths (or colors) of light. As such, the lens is really focusing a number of images at varying distances from the lens, one for each wavelength present in the incident light, which creates an unfocused overall image.

To combat chromatic aberration, you can place light filter elements (if they're available to you) over the magnifying lens. These filters block out certain wavelengths, so you may find that you can affect the rainbow colors. You can also stack these filters to block out multiple wavelengths simultaneously.

Higher-quality lenses may address chromatic aberration by adding two specific lenses in contact with each other. These lenses' respective dispersive powers compensate for the chromatic aberration of the original lens.

Chapter 23

Ten Major Optics Discoveries — and the People Who Made them Possible

• •

In This Chapter

▶ Introducing some of the key scientists in optics's history

▶ Examining the experiments and discoveries that made optics what it is

• •

*E*ach generation of scientists stands on the shoulders of those who have come before them, with every new discovery being built out of the existing pool of knowledge. The study of optics is no different. In this chapter, you find out about some of the key discoveries that have moved the field of optics along the path toward where it is today.

The Telescope (1610)

Galileo Galilei's intellectual offspring are many. He has been called "the father of science" and "the father of modern physics." However, his work with the telescope is what earned him the label "the father of modern observational astronomy."

Though Galileo didn't invent the first telescope and isn't even the first to have looked into space with one, he's credited with making it into an instrument for the systematic study of scientific questions. His 1610 book *Starry Messenger* described how he had discovered the moons of Jupiter by using the device.

The original telescope Galileo designed had a magnification of about 3 times, but his improvements over the years resulted in a telescope with a magnification of 30 times. His observations eventually led him to understand that the geocentric (earth-centered) model of the universe was incorrect, eventually bringing him to a heresy trial by the Catholic Church, which resulted in him being confined to house arrest for the remainder of his life.

Optical Physics (Late 1600s)

Although Galileo got the ball rolling on a study of the telescope, Sir Isaac Newton was the guy who successfully designed the first telescope that used mirrors rather than lenses, as I describe in Chapter 19. But this instrument was far from Newton's most significant contribution to the field of optics. In fact, until Newton, optics really wasn't a field at all.

Newton's research into optics was the most extensive of the time, resulting in two books on the subject: the 1675 *Hypothesis of Light* and the 1704 *Opticks*.

Newton's discoveries and experiments involving optics include the following:

- He used a prism to decompose white light into a spectrum of colors and recomposed the spectrum back into white light by using a lens and a second prism.

- Shining different types of light on different-colored objects, he demonstrated that color is a property of light, not something created by the material itself.

- Based on the color breakthrough, Newton discovered *chromatic aberration,* which is where lenses cause slight dispersion of light into colors (head to Chapter 7). He invented the *Newtonian telescope,* the first reflecting telescope, which didn't suffer from this flaw.

- He developed a particle theory of light (see Chapter 3).

Diffraction and the Wave Theory of Light (Late 1600s)

Though Newton certainly went the farthest in developing a comprehensive theory of optics, his wasn't the only game in town. (And by "town," I mean 17th-century Europe.)

The Dutch mathematician and astronomer Christiaan Huygens, a slightly older contemporary of Newton, had a very different take on light from his esteemed English counterpart. Huygens supported the idea that light traveled in waves rather than as particles.

The problem is that although a particle theory is mathematically simpler — the light rays are just treated as straight lines, after all — some phenomena, such as diffraction, simply made more sense with Huygens's wave theory than Newton's particle one. (This approach to treating light as a wave is covered in Chapters 2 and 3.)

At the time, though, Newton was such a powerful voice in physics that he drowned out Huygens. Most scientists used Newton's particle theory, at least until they encountered a compelling reason for them to change (see the following section).

Two-Slit Experiment (Early 1800s)

For about a century, Newton's particle theory held sway over Huygens's wave interpretation of light, but that changed in the early 19th century when Thomas Young performed his two-slit experiment.

The two-slit experiment is described in Chapter 11, so I keep it brief here. Young found that when you send a beam of light through two small slits, you get a series of many light and dark bands instead of getting two bands of light (which is what you'd expect from a particle theory). This series is an *interference pattern,* the sort of thing that can only be explained by a wave theory of light such as the one developed by Christiaan Huygens over a century earlier (see the preceding section).

Polarization (Early 1800s)

In the years following Young's two-slit experiment (see the preceding section), other experiments continued to call into question the particle theory of light and lend support to the wave theory. One of the leaders in this movement was French physicist Augustin-Jean Fresnel, who performed experiments in polarization (see Chapter 8) and also followed up on Huygens's work in diffraction (see Chapter 12).

Fresnel's work expanded on and refined Huygens' concepts of waves and wavefronts tremendously in the 19th century, so much so that the fundamental insight into this approach is known as the *Huygens-Fresnel principle,* which is the principle that you can treat each point in a wave as the source of a new wave. (Chapter 2 covers this use of wavefronts more fully.)

Rayleigh Scattering (Late 1800s)

Ever wonder why the sky is blue? This question got its answer in the late 1800s when John William Strutt, 3rd Baron Rayleigh, discovered an effect known as *Rayleigh scattering.*

Small particles, such as the molecules that make up the air, or other atoms, can scatter light waves. The scattered sunlight in the atmosphere creates a blue backdrop when you look into the sky. (Head to Chapter 4 for more on Rayleigh scattering.)

Lord Rayleigh received a 1904 Nobel Prize, but not for this achievement! His Nobel was actually for his co-discovery of the element argon.

Electromagnetics (1861)

The first half of the 19th century brought about an amazing understanding of the fundamental connection between electrical and magnetic forces, which had previously been thought completely different. By the end of the century, scientists understood that these two forces were different aspects of the same electromagnetic force and that electromagnetic radiation was different frequencies of "invisible light." This key insight has helped to define optics ever since.

Much of the experimental work that laid the foundation for this understanding was performed by Michael Faraday. Faraday had little in the way of formal schooling, so James Clerk Maxwell transformed Faraday's discoveries into the mathematical form eventually known as "Maxwell's equations," which Maxwell published in 1861.

Electro-Optics (1875 and 1893)

Electromagnetic fields can change the refractive index of materials, as demonstrated in the late 1800s by the discovery of electro-optic effects. (Chapter 20 covers electro-optics in greater detail.) In 1875, John Kerr discovered the first electro-optic effect, aptly named the *Kerr effect,* which indicated that the change in refractive index is proportional to the square of the electric field applied.

Several years later, Friedrich Carl Alwin Pockels discovered that certain forms of crystals possessed this property as well, although the change in refractive index was proportional to the first power of the electric field (instead of squaring it). This case is known as the *Pockels effect.*

Photon Theory of Light (1905)

Albert Einstein is known by many for his theory of relativity, but the same year that he introduced that concept, he also wrote another paper, which helped to reconcile the conflict that had come up two and a half centuries earlier between Newton and Huygens: Was light a wave or a particle?

For a century, the wave theory had been winning out, but with one big problem. A wave required something to be doing the waving. Scientists had long theorized that some strange substance, called the *ether,* was the medium that light waves disturbed (just like water waves disturb water). The problem was that all attempts to find the ether had categorically failed.

Einstein wasn't really thinking about the ether when he tackled the problem of explaining the photoelectric effect (from Chapter 3), where shining a light on a metal plate could cause the release of electrons. His explanation involved merging the wave and particle theories of light into one unified model, which he called *quanta* after a similar idea proposed by Max Planck (to solve an entirely different problem) in 1900. Einstein's quanta later became known as the photon.

The trick with the photon is understanding that it's a little packet of energy that has a wavelength and frequency. In some cases, thinking of light as a wave rather than a particle is better, and vice versa, but neither is more correct than the other.

Not only did Einstein's discovery do the job of explaining the photoelectric effect, but it (along with Planck's earlier work) had the unintended consequence of ushering in the era of quantum physics, with this new wavelike particle of light — the photon — right at the center of it. No ether required.

The Maser (1953) and The Laser (1960)

The principles of quantum physics, as they pertain to optics, are never more fully manifested than in that wonder of modern science: the laser.

Chapter 14 describes the science of the laser, but the original laser isn't at all what you probably picture when you hear the word today. Rather than a "light amplification by stimulated emission of radiation" (laser) system, the original was a *maser* because it emitted only microwave radiation, not visible light. (These names were actually coined in 1959.)

Charles Hard Townes, who had invented the maser (and, along with some Russians who did similar work at the same time, received the 1964 Nobel Prize in Physics for it), later worked together with Arthur Leonard Schawlow to develop the theory of the optical laser.

Today, the laser is one of the most ubiquitous devices around. You use lasers to check out items at the grocery store, create and view holographic images, listen to music and watch movies, store computer data, do scientific research, target weapons, perform medical procedures, cut and weld metal, and point at things on a screen during lectures. (Flip to the chapters in Part V for more on many of these applications.)

In optics, one of the most important results of the laser has been the ability to use it to further the understanding of how light behaves in the quantum world. An entire field of physics, quantum optics, seeks to understand the quantum behavior of light. Physicists have used the laser to return to Young's two-slit experiment (see the earlier section "Two-Slit Experiment (Early 1800s)") and discovered, amazingly, that the wave interpretation can be applied even when only one photon at a time is going through the slits!

Index

Apple & Macs

iPad For Dummies
978-0-470-58027-1

iPhone For Dummies,
4th Edition
978-0-470-87870-5

MacBook For Dummies, 3rd
Edition
978-0-470-76918-8

Mac OS X Snow Leopard For
Dummies
978-0-470-43543-4

Business

Bookkeeping For Dummies
978-0-7645-9848-7

Job Interviews
For Dummies,
3rd Edition
978-0-470-17748-8

Resumes For Dummies,
5th Edition
978-0-470-08037-5

Starting an
Online Business
For Dummies,
6th Edition
978-0-470-60210-2

Stock Investing
For Dummies,
3rd Edition
978-0-470-40114-9

Successful
Time Management
For Dummies
978-0-470-29034-7

Computer Hardware

BlackBerry
For Dummies,
4th Edition
978-0-470-60700-8

Computers For Seniors
For Dummies,
2nd Edition
978-0-470-53483-0

PCs For Dummies,
Windows
7th Edition
978-0-470-46542-4

Laptops For Dummies,
4th Edition
978-0-470-57829-2

Cooking & Entertaining

Cooking Basics
For Dummies,
3rd Edition
978-0-7645-7206-7

Wine For Dummies,
4th Edition
978-0-470-04579-4

Diet & Nutrition

Dieting For Dummies,
2nd Edition
978-0-7645-4149-0

Nutrition For Dummies,
4th Edition
978-0-471-79868-2

Weight Training
For Dummies,
3rd Edition
978-0-471-76845-6

Digital Photography

Digital SLR Cameras &
Photography For Dummies,
3rd Edition
978-0-470-46606-3

Photoshop Elements 8
For Dummies
978-0-470-52967-6

Gardening

Gardening Basics
For Dummies
978-0-470-03749-2

Organic Gardening
For Dummies,
2nd Edition
978-0-470-43067-5

Green/Sustainable

Raising Chickens
For Dummies
978-0-470-46544-8

Green Cleaning
For Dummies
978-0-470-39106-8

Health

Diabetes For Dummies,
3rd Edition
978-0-470-27086-8

Food Allergies
For Dummies
978-0-470-09584-3

Living Gluten-Free
For Dummies,
2nd Edition
978-0-470-58589-4

Hobbies/General

Chess For Dummies,
2nd Edition
978-0-7645-8404-6

Drawing
Cartoons & Comics
For Dummies
978-0-470-42683-8

Knitting For Dummies,
2nd Edition
978-0-470-28747-7

Organizing
For Dummies
978-0-7645-5300-4

Su Doku For Dummies
978-0-470-01892-7

Home Improvement

Home Maintenance
For Dummies,
2nd Edition
978-0-470-43063-7

Home Theater
For Dummies,
3rd Edition
978-0-470-41189-6

Living the
Country Lifestyle
All-in-One
For Dummies
978-0-470-43061-3

Solar Power Your Home
For Dummies,
2nd Edition
978-0-470-59678-4

Internet

Blogging For Dummies,
3rd Edition
978-0-470-61996-4

eBay For Dummies,
6th Edition
978-0-470-49741-8

Facebook For Dummies,
3rd Edition
978-0-470-87804-0

Web Marketing
For Dummies,
2nd Edition
978-0-470-37181-7

WordPress
For Dummies,
3rd Edition
978-0-470-59274-8

Language & Foreign Language

French For Dummies
978-0-7645-5193-2

Italian Phrases
For Dummies
978-0-7645-7203-6

Spanish For Dummies,
2nd Edition
978-0-470-87855-2

Spanish
For Dummies,
Audio Set
978-0-470-09585-0

Math & Science

Algebra I
For Dummies,
2nd Edition
978-0-470-55964-2

Biology For Dummies,
2nd Edition
978-0-470-59875-7

Calculus For Dummies
978-0-7645-2498-1

Chemistry For Dummies
978-0-7645-5430-8

Microsoft Office

Excel 2010 For Dummies
978-0-470-48953-6

Office 2010 All-in-One
For Dummies
978-0-470-49748-7

Office 2010 For Dummies,
Book + DVD Bundle
978-0-470-62698-6

Word 2010 For Dummies
978-0-470-48772-3

Music

Guitar For Dummies,
2nd Edition
978-0-7645-9904-0

iPod & iTunes For
Dummies, 8th Edition
978-0-470-87871-2

Piano Exercises
For Dummies
978-0-470-38765-8

Parenting & Education

Parenting For Dummies,
2nd Edition
978-0-7645-5418-6

Type 1 Diabetes
For Dummies
978-0-470-17811-9

Pets

Cats For Dummies,
2nd Edition
978-0-7645-5275-5

Dog Training For Dummies,
3rd Edition
978-0-470-60029-0

Puppies For Dummies,
2nd Edition
978-0-470-03717-1

Religion & Inspiration

The Bible For Dummies
978-0-7645-5296-0

Catholicism For Dummies
978-0-7645-5391-2

Women in the Bible
For Dummies
978-0-7645-8475-6

Self-Help & Relationship

Anger Management
For Dummies
978-0-470-03715-7

Overcoming Anxiety
For Dummies,
2nd Edition
978-0-470-57441-6

Sports

Baseball
For Dummies,
3rd Edition
978-0-7645-7537-2

Basketball
For Dummies,
2nd Edition
978-0-7645-5248-9

Golf For Dummies,
3rd Edition
978-0-471-76871-5

Web Development

Web Design
All-in-One
For Dummies
978-0-470-41796-6

Web Sites
Do-It-Yourself
For Dummies,
2nd Edition
978-0-470-56520-9

Windows 7

Windows 7
For Dummies
978-0-470-49743-2

Windows 7
For Dummies,
Book + DVD Bundle
978-0-470-52398-8

Windows 7 All-in-One
For Dummies
978-0-470-48763-1

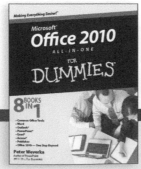

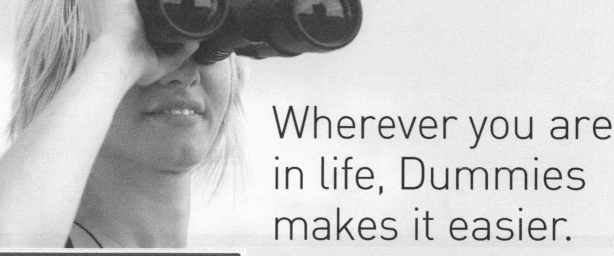

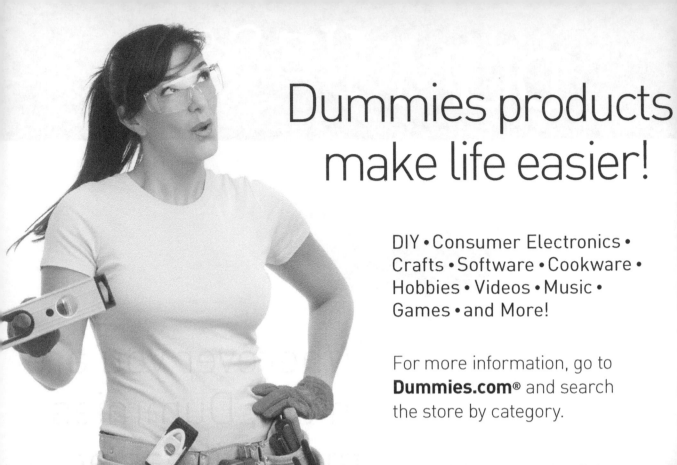

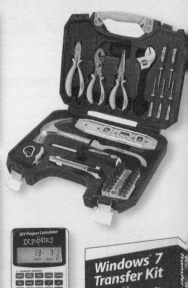